普通高等教育"十二五"规划教材

Bilingual Course of Biohydrometallurgy for Nonferrous Metals
有色金属生物冶金

薛济来　主编

北　京
冶金工业出版社
2012

内 容 提 要

有色金属生物冶金主要采用生物技术来对矿物中的有色金属进行富集、分离、提取和回收利用，通常由微生物来进行矿石的细菌氧化或生物氧化。生物冶金工艺成本低、污染少、资源利用率高，目前已应用于难处理金矿、铜硫化矿等。生物冶金具有超过常规处理贫矿的技术优势，正发展成为国际上有色冶金研究的热点之一。

本书主要内容包括：有色金属生物冶金技术基础、冶金方法、工艺特征和工业应用（提取金、铜、镍、锌）。文中均加以简要注释，以方便自学。书后附相关专业词汇。

本书适用于有色冶金、冶金工程、工业生态、矿物工程、环境工程等专业的高年级本科生和研究生科研和双语教学，也可供研发人员和生产技术人员参考。

图书在版编目(CIP)数据

有色金属生物冶金：中文、英文/薛济来主编. —北京：冶金工业出版社，2012.8

普通高等教育"十二五"规划教材
ISBN 978-7-5024-6153-9

Ⅰ.①有… Ⅱ.①薛… Ⅲ.①细菌冶金—高等学校—教材—汉、英 Ⅳ.①TF18

中国版本图书馆 CIP 数据核字 (2012) 第 309161 号

出 版 人　谭学余
地　　址　北京北河沿大街嵩祝院北巷39号，邮编100009
电　　话　(010)64027926　电子信箱　yjcbs@cnmip.com.cn
责任编辑　刘小峰　美术编辑　李 新　版式设计　孙跃红
责任校对　禹 蕊　责任印制　李玉山

ISBN 978-7-5024-6153-9

冶金工业出版社出版发行；各地新华书店经销；北京百善印刷厂印刷
2012年8月第1版，2012年8月第1次印刷
787mm×1092mm　1/16；8.5印张；201千字；123页
28.00元

冶金工业出版社投稿电话: (010)64027932　投稿信箱: tougao@cnmip.com.cn
冶金工业出版社发行部　电话: (010)64044283　传真: (010)64027893
冶金书店　地址: 北京东四西大街46号(100010)　电话: (010)65289081(兼传真)
(本书如有印装质量问题，本社发行部负责退换)

前　言

有色金属在国民经济和国防事业中占有重要的战略地位，是实现工业化、城镇化和信息化的基础支撑材料。当前我国有色金属产业快速发展，同时又面临着资源短缺、环境污染等瓶颈问题。利用生物技术处理低品位矿物和开发环境友好的有色金属提取工艺，是近年来国际上兴起和迅速发展的前沿技术。通过英文文献载体，可直接交流、快捷获取国际上有关这一领域的发展趋势和技术动态的最新信息，有助于从事相关领域的研究开发和专业教育工作。

有色金属生物冶金主要涉及利用细菌和生物技术对矿物和废弃物中所含有色金属进行富集、分离、提取和回收利用。本书内容包括：生物冶金物理化学基础、有色金属生物冶金方法、工业应用（提取铜、金、钴、镍、锌等有色金属）及二次资源综合利用。书中内容主要参考国外有关英文专著和学术期刊，力求保持其英文原貌，并加以简要注释，以方便自修。

本书选编时着重对有关基本原理、方法、典型技术应用的精要介绍，同时兼顾基础性和应用性，可广泛适用于冶金工程、有色冶金、化工、材料、矿物加工、环境和资源工程、生物技术、应用化学等专业的大学高年级本科生、研究生、中高层研发人员和管理人员提升专业知识和当代科技英语水平两方面的需求。

作者多年在北京科技大学冶金与生态工程学院讲授有色金属生物冶金英中双语课程，曾使用过本书内容并在教学过程中获益匪浅。李想、冯鲁兴、张亚楠在本书编辑过程中付出了辛勤劳动，在此一并感谢。

作者深感水平有限，研究不足，书中难免遗漏不当之处，敬希业内同仁和读者指正。

薛济来
2012 年 8 月于北京

Preface

Nonferrous metals play a strategic important role in national economics and defense, which are basic materials supporting developments of industrialization, urbanization and informationization in China for today. The Chinese nonferrous metals industry has made great progress in recent years, while at the same time it faces challenges in resource shortage and environmental pollution. Biotechnology treating low-grade metals ores and creating environment-friendly process to extract nonferrous metals are the internationally emerging and fast developing cut-edge technology. Through English technical literature, you can obtain the most-updated knowledge about the international trend of development in your interesting fields and improve your professional skills and research work.

Biohydrometallurgy for nonferrous metals is dealt with the extracting, concentrating, separating and recycling of nonferrous metals from minerals and spend residues using bacteria and biotechnology. This book contains fundamentals in biotechnology and bacteria, biohydrometallugical processing methods for nonferrous metals, their industrial implementations (extraction of Cu, Au, Co, Ni and Zn metals) and metals recovery from secondary sources. These contents have been referred to scientific books and professional journals in English origin, along with providing explanation notes to assist readers for self-study.

In preparing this book, emphases have been placed on supplying information on fundamentals, processing methods and some typical industrial applications in biohydrometallurgy for nonferrous metals, where a balance is properly made between the technical theory and the industrial reality. The book is intended for senior students, postgraduates and young and middle age professionals in many different fields, such as metallurgical engineering, nonferrous metallurgy, chemical engineering, materials technology, minerals engineering, environmental and resource engineering, biotechnology, applied chemistry, etc. It may serve the purpose of improving both their scientific knowledge and contemporary English skill.

The author has used the same contents as mentioned above for several years in teaching bilingual course at School of Metallurgical and Ecological Engineering, University of Science and Technology Beijing, and benefited the most from a good interaction between the learning and the teaching during that time. Last but not the least, the author would like to thank Li Xiang, Feng Luxing, and Zhang Yanan for their assistance in preparing this manuscript, and also invite the readers and colleagues to provide their comments and criticism on this edition.

Xue Jilai

Beijing, August 2012

Content

1 Biohydrometallurgy and Bacteria ··· 1

1.1 Introduction ·· 1
1.2 Development in Hydrometallurgy ··· 2
1.3 Bacteria and Phylogeny ··· 4
1.4 Nutrition ·· 7
   1.4.1 Carbon nutrition sources ·· 7
   1.4.2 Nitrogen nutrition sources ·· 7
1.5 Energy Sources ··· 8
   1.5.1 Iron oxidation ·· 8
   1.5.2 Sulfur oxidation ·· 9
   1.5.3 Electron donors and electron acceptors ···································· 10
   1.5.4 Mineral oxidating ability ·· 11
1.6 Mesophilic and Acidophiles in Mineral Bioleaching ····························· 12
1.7 Summary ··· 15
Questions ·· 15
Note ·· 15

2 Surface Chemistry of Bacterial Leaching ·· 17

2.1 Introduction ·· 17
2.2 Adhesion and Attachment of Bacteria to Mineral Surfaces ····················· 18
2.3 Theory of Bacterial Attachment to Surfaces ······································· 20
2.4 Biofilm Colloid Formation on Bacterial Cells ···································· 26
2.5 Summary ··· 27
Questions ·· 27
Note ·· 27

3 Electrochemistry of Mineral Dissolution and Bioleaching ···················· 29

3.1 Introduction ·· 29
3.2 Electrochemical Mechanism of Oxidative and Reductive Reactions ·········· 30
   3.2.1 Anodic and cathodic reactions ··· 30
   3.2.2 Influence of the electronic structure of the mineral on the dissolution rate ··· 31
3.3 Applications of Electrochemical Mechanism in Leaching ······················ 32

    3.3.1   Chemical leaching of pyrite by ferric ions ·· 32
    3.3.2   Effect of bacterial action on the mixed potential of pyrite ·················· 33
    3.3.3   Effect of electrochemical bioleaching on the copper recovery ············ 35
3.4   Electrochemical Kinetics and Modeling ·· 35
    3.4.1   Kinetics of the oxidative dissolution of sphalerite ······························· 35
    3.4.2   Electrochemical kinetics and model of bacterial leaching ··················· 37
3.5   Summary ·· 39
Questions ··· 39
Note ·· 40

4 Biohydrometallurgy of Copper ·· 41

4.1   Introduction ··· 41
4.2   Definitions and Mineralogy Related to Copper Leaching ···················· 41
    4.2.1   Pyrite ·· 42
    4.2.2   Secondary sulfides ·· 42
    4.2.3   Primary sulfides ··· 43
4.3   Physico-Chemical Leaching Variables ·· 43
    4.3.1   Surface area ·· 43
    4.3.2   Acid levels ·· 44
    4.3.3   Oxidants ··· 44
    4.3.4   Agglomeration ··· 44
    4.3.5   Curing time ·· 45
    4.3.6   Permeability ··· 45
4.4   Bacterial Leaching Variables ·· 45
    4.4.1   Acidity ··· 45
    4.4.2   Oxygen ··· 46
    4.4.3   Nutrition ·· 46
    4.4.4   Heat ·· 47
    4.4.5   Mineralogy ·· 47
    4.4.6   Bacterial inoculation ·· 47
    4.4.7   Iron ··· 48
4.5   Heap Operating Variables ··· 48
    4.5.1   Irrigation distribution ··· 48
    4.5.2   Solution stacking ··· 49
    4.5.3   Solution collection ··· 49
    4.5.4   Pad stacking/configuration ··· 50
4.6   Leach Solution Processing ··· 50
    4.6.1   Copper cementation ··· 51
    4.6.2   Direct electrowinning ·· 51

4.6.3	Solvent extraction	51
4.6.4	Electrowinning	53

4.7 Commercial Installations and Environmental Considerations · · · · · · · · · · · 54

    4.7.1 In situ leaching · · · · · · · · · · · 55

    4.7.2 Dump leaching · · · · · · · · · · · 56

    4.7.3 Heap leaching · · · · · · · · · · · 56

4.8 Summary · · · · · · · · · · · 58

Questions · · · · · · · · · · · 58

Note · · · · · · · · · · · 58

5 Biooxidation of Gold-Bearing Ores · · · · · · · · · · · 60

5.1 Introduction · · · · · · · · · · · 60

5.2 BIOX® Bacterial Culture · · · · · · · · · · · 61

5.3 Chemical Reactions and Process Control · · · · · · · · · · · 63

    5.3.1 Chemical reactions · · · · · · · · · · · 63

    5.3.2 Influence of ore mineralogy · · · · · · · · · · · 64

    5.3.3 Effect of temperature and cooling requirements · · · · · · · · · · · 65

    5.3.4 pH control · · · · · · · · · · · 66

    5.3.5 Oxygen supply · · · · · · · · · · · 67

5.4 Operations Conditions and Process Requirements · · · · · · · · · · · 68

    5.4.1 Bioreactor configuration · · · · · · · · · · · 68

    5.4.2 Rate of sulfide mineral oxidation and gold dissolution · · · · · · · · · · · 69

    5.4.3 General process requirements · · · · · · · · · · · 70

    5.4.4 Effects of chloride and arsenic on BIOX® process · · · · · · · · · · · 71

5.5 Summary · · · · · · · · · · · 72

Questions · · · · · · · · · · · 73

Note · · · · · · · · · · · 73

6 Biohydrometallurgical Processing of Colt, Nickel and Zinc · · · · · · · · · · · 74

6.1 Introduction · · · · · · · · · · · 74

6.2 Cobalt Bioleaching with Autotrophic and Mixotrophic Bacteria · · · · · · · · · · · 75

6.3 Nickel Bioleaching with Autotrophic and Mixotrophic Bacteria · · · · · · · · · · · 77

6.4 Zinc Bioleaching with Autotrophic and Mixotrophic Bacteria · · · · · · · · · · · 78

6.5 Metal Mobilization by Microbially Generated Acids/Ligands · · · · · · · · · · · 80

6.6 Summary · · · · · · · · · · · 84

Questions · · · · · · · · · · · 84

Note · · · · · · · · · · · 85

7 Biohydrometallurgical Recovery of Heavy Metals from Industrial Wastes ········ 86

7.1 Introduction ········ 86
7.2 Metal Recovery from Waste Sludge and Fly Ash ········ 87
7.3 Metal Recovery from Mine Waste and Nuclear Waste ········ 89
7.4 Metal Recovery from River Sediments and Metal Finishing Waste Water ········ 92
7.5 Summary ········ 94
Questions ········ 94
Note ········ 95

8 Biohydrometallurgical Recovery of Value Metals from Secondary Sources ········ 96

8.1 Introduction ········ 96
8.2 Metal Recovery from Electronic Wastes ········ 96
8.3 Metal Recovery from Battery Wastes ········ 100
8.4 Metal Recovery from Spent Petroleum Refinery Catalyst ········ 103
8.5 Summary ········ 105
Questions ········ 106
Note ········ 106

Appendix ········ 107

References ········ 122

1 Biohydrometallurgy and Bacteria

本章要点 本章简要介绍了生物冶金技术近年来的研究和发展。对微生物、不同温度范围的细菌种类、细菌营养、矿物浸出过程、生物浸出过程、生物冶金技术在有色金属提取冶金中应用的有关概念、技术背景和国内外产业化实例做了概述。

1.1 Introduction

Natural bioleaching has been taking place for almost as long as the history of the world, but it is only in the last few decades that we have realized that bioleaching is responsible for acid production in some mining wastes, and that this bacterial activity can be used to liberate some metals. Today large-scale, biohydrometallurgy facilities extract copper and enhance gold recovery from ores and concentrates, and the commercially applied environmental biotechnologies remediate metal-contaminated waters and degrade cyanide. These commercial operations demonstrate the qualities robust nature, operational simplicity, health, safety and environmental benefits, capital and operating cost advantages and improved performance that have made bioleaching/biooxidation and bioremediation viable process options for the mining and nonferrous metals industry. With commercial success, the industry now anticipates better efficiency from existing biotechnical processes, improvements in applications and novel processes to further enhance the utility and extent of application.

The low pH, metal-rich, inorganic mineral environment in which bioleaching reactions occur is populated by a group of bacteria which are highly adapted to growth under these conditions. The bacteria most commonly isolated from inorganic mining environments are *Thiobacillus ferrooxidans*, *Thiobacillus thiooxidans* and *Leptospirillum ferrooxidans*. These bacteria are considered to be the most important in most industrial leaching processes. The recently described moderately thermophilic bacterium, *Thiobacillus caldus*, which grows optimally at 45°C also grows well at between 30°C and 40°C and may be readily isolated from bioleaching processes which operate at this temperature. From a physiological viewpoint this bacterium is the moderately thermophilic equivalent of *Thiobacillus thiooxidans* and the role of *T. caldus* in many industrial operations is almost certainly greater than that has been generally recognized. A number of species of heterotrophs which grow in very close association with the obligate autotrophs have been found, most of which belong to the genus *Acidiphilium*. Other acidophilic bacteria which have been reported from leaching environments include the facultative heterotrophic bacteria *Thiobacillus acidophilus* and *Thiobacillus coprinus*.

This chapter presents a brief description of development in biohydrometallurgy, some fundamentals knowledge about bacteria, nutrition and energy sources, and the extent and scale of commercial biotechnology applications in processing minerals containing nonferrous metals.

1.2 Development in Hydrometallurgy

Although the terms bioleaching and biooxidation are often used interchangeably, there are distinct technical differences between these process technologies. Bioleaching refers to the use of bacteria, principally *Thiobacillus ferrooxidans*, *Leptospirillum ferrooxidans* and thermophilic species of *Sulfobacillus*, *Acidianus* and *Sulfolobus*, to leach a metal of value such as copper, zinc, uranium, nickel and cobalt from a sulfide mineral. Bioleaching places the metal values of interest in the solution phase during oxidation. These solutions are handled for maximum metal recovery and the solid residue is discarded.❶ Minerals biooxidation refers to a pretreatment process that uses the same bacteria as bioleaching to catalyze the degradation of mineral sulfides,❷ usually pyrite or arsenopyrite, which host or occlude gold, silver or both. Biooxidation leaves the metal values in the solid phase and the solution is discarded.

The discovery in 1947 that bacteria are associated with acid rock drainage and the characterization and naming of *Thiobacillus ferrooxidans* in 1951 spurred research on the role of these organisms in oxidizing mineral sulfides. Research published in the 1950s led to the 1958 patent that preceded the industrial scale application of copper dump leaching by Kennecott Copper Corporation in the early 1960s. Research and development between 1960 and 1980 yielded important information:

(1) Metabolic pathways for sulfur and iron oxidation in thiobacilli were described;

(2) Morphological and structural characteristics of the leaching bacteria identified;

(3) Genetics of metal-microbe interactions introduced;

(4) Moderate and extreme thermophiles that oxidize reduced sulfur and iron compounds and mineral sulfides discovered and partially characterized;

(5) Microbial-mineral interactions and mineral metabolism described;

(6) Metal, iron concentration, nutrient, light, pressure, aeration, and temperature, affects on bioleaching quantified;

(7) Extent of microbial leaching of mineral sulfide ores and concentrates (for example, pyrite in coal, arsenic sulfides, chalcopyrite, and complex mineral sulfides) defined;

(8) Ecology of copper dump leach operations studied;

(9) Heterotrophic microbial processes of mineral's leaching surveyed;

(10) Bioleaching of mineral sulfide concentrates in continuous, stirred-tank reactors perfected;

(11) Large-scale test facilities to evaluate copper bioleaching employed.

The results of these studies were chronicled in the International Biohydrometallurgy Symposia of the 1970s, an excellent book on thermophilic microorganisms, and hundreds of journal publications.

The 1990s has been the decade of commercial awakening of bioleaching/biooxidation. Twelve of the world's fourteen commercially-operating facilities were commissioned in the first five years of this decade. This commercial activity sparked notable findings reported in an increasing number of journal publications and symposia devoted to the discipline. Research and development were made in the biochemistry, genetics and molecular aspects of the leaching organisms and-at the other end of the spectrum-test work requirements, engineering considerations and economics of tank and heap bioleaching/biooxidation scale-up. It is somehow paradoxical that, despite the considerable advances that have been made over these years, a fundamental aspect-the relative importance and contribution of direct and indirect leaching-eludes scientists and practitioners of the technology.

Commercial applications of aerated, stirred tanks for refractory-sulfidic gold concentrates and bioheaps for chalcocite and refractory-sulfidic gold ores have shifted the focus of R&D somewhat to topics that will improve these processes, diminish capital and operating costs and extend the process applications. Achievements made in this context are:

(1) Improved aeration systems for stirred-tanks that will improve utilization efficiency, and decrease operating costs;

(2) Aerated, stirred-tank and bacterial ferric generation systems for chalcopyrite concentrate bioleaching developments that promise significant cost reductions over smelting and could open new mines in remote locations;

(3) Aerated, stirred tanks for Co, Zn and other mineral sulfide bioleaching and for recovery of less-common metals (e.g., In);

(4) Bioheap leaching of concentrates, which would reduce capital and operating costs for processing concentrates;

(5) Techniques to decrease cyanide consumption of biooxidized residues, reducing operating costs of gold plants;

(6) Innovative bioheap technologies for diverse climates (e.g., high rainfall areas, tropics, Arctic regions), disparate ore types (e.g., clayey ores), and enhanced aeration;

(7) Effective application of extremely thermophilic bacteria in stirred-tank reactors and bioheaps;

(8) Better understanding of solution chemistry equilibria in bioheaps to improve kinetics and minimize leach solution treatment costs.

Biooxidation/bioleaching research that promises to make a difference from a fundamentals perspective encompasses:

(1) Bioleaching of chalcopyrite ore to allow low-cost heap and in situ leaching of deposits;

(2) Genetically altered bacteria which improve biooxidation/bioleaching and can electively compete with native bacteria in commercial plants;

(3) Definitive understanding of the direct versus indirect attack and the relative importance to stirred tank and bioheap reactors.

In China, the biohydrometallurgy has been also developed in recent years and has been applied to a large extent in industrial productions. Zijinshan Gold & Copper Mine is situated 14.6km north of Shanghang county in Fujian province, along the left band of the beautiful and picturesque Tingjiang river (Fig. 1.1). It has extremely rich gold and copper reserves. According to historical records, people have panned for gold here since the Sung Dynasty (around 1040 AD). Zijinshan Mine is one of the large non-ferrous metal mines being discovered in the 1980s in China. It has a typical vertical zoning with gold in the upper zone and copper underneath, which is a super-large scale porphyry-type ore deposits. Gold deposited at 600-1100m elevation within the oxidation zone, while copper sulfide ores below the 600-700m elevation. Zijinshan Gold & Copper Mine has been to be a large scale successful example of gold heap leaching in this humid and rainy environment in southern China. Furthermore, Zijinshan copper mine is the largest chalcocite deposit in China. The attempt to recover copper from the ore by bioleaching began at the end of 1998. Following the metallurgical studies carried out over 2 years, a pilot plant consisting of bio-heap leaching and SX–EW was built at the Zijinshan copper mine at the end of 2000, with a production capacity of 300t/a copper cathode. After its successful operation for 1.5 years, the plant was scaled up to a capacity of 1,000t/a copper cathode by june 2002. In 2005, a commercial bio-heap leach plant with a capacity of 10,000t/a was built.

Fig. 1.1 Zijin opencaste gold mining

In order to improve research and technology in biohydrometallurgy in China, two state key laboratories have been built in recent years. One is the Key Laboratory of Biometallurgy of Ministry of Education in Central South University and another is National Engineering Laboratory of Biometallurgy supported by General Research Institute for Nonferrous Metals.

1.3 Bacteria and Phylogeny

T. ferrooxidans, *T. thiooxidans* and *L. ferroxidans* are used in bioleaching process, which are all autotrophic and acidophilic bacteria. Both *T. ferrooxidans* and *T. thiooxidans* are rod-shaped while *L. ferrooxidans* is spiral-shape, as shown in Fig. 1.2. *T. ferrooxidans* is able to use ferrous iron or reduced sulfur compounds as an electron donor, while *T. thiooxidans* is capable of using only reduced sulfur compounds and *L. ferrooxidans*, only ferrous iron. All three bacteria use oxygen as a terminal electron acceptor, although *T. ferrooxidans* is able to grow using ferric iron as an electron acceptor provided reduced sulfur compounds are available to serve as an electron donor.

T. ferrooxidans has been described as a species of convenience. Based upon DNA IA homology studies, strains called *T. ferrooxidans* were placed by Harrison into as many as seven subgroups. Five *T. ferrooxidans* homology subgroups had G+C contents of 56%-59%, one subgroup was shown to be an *L. ferrooxidans* isolate and another, represented by a single isolate (*T. ferrooxidans* m-1) had a G+C content of 65% and has not been shown to oxidize sulfur. Therefore only five of the subgroups can be considered to be *T. ferrooxidans*.

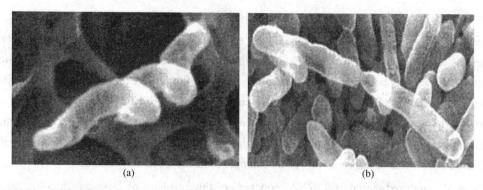

Fig. 1.2   A scanning electron microscope image of (a) *L. ferrooxidans* DSM2705 (×18000) and (b) *T. ferrooxidans* ATCC33020 (×15000)

T. thiooxidans may also consist of more than one grouping with most strains having a G+C content of 52%-53%. An exception is strain DSM612 which has a 62% G+C content and may be more similar to *Thiobacillus albertis*. Another strain has a G+C content of 58%, which falls within the range for *T. ferrooxidans*. A difficulty in using the ability to oxidize iron or sulfur as a means of distinguishing *T. ferrooxidans* from *T. thiooxidans* is that many *T. ferrooxidans* strains may exhibit a delay in switching between iron with the result that researchers may assume that the organism is incapable of oxidizing one or other energy source.

The taxonomic status of the genus *Leptospirillum* remains to be fully resolved. There is no valid description of the organism in Bergey's Manual of Determinative Bacteriology, although reference is made to *Leptospirillum* in the section on the genus *Thiobacillus*. Acidiophilic, obligate autotrophic bacteria which grow on ferrous iron and sulfide minerals and which have a helically curved rod-shaped morphology are considered to be strains of *Leptospirillum*. The spiral forms can be of varying length and generally have a long polar flagellum. The bacterium was originally described by Markosyan from samples collected in Armenia but there is evidence that this *L. ferrooxidans* L15 strain may be atypical. Harrison and Norris compared protein electrophoretic patterns, DNA-DNA homology and mol% G+C ratios for *L. ferrooxidans* and several *Leptospirillum*-like bacteria. They found that the bacteria could be divided into two groups with a G+C content of either 51%-52% or 55%-56mol%. Bacteria which fit the description of *L. ferrooxidans* have been isolated from numerous environments including water samples from uranium mines in Canada and Mexico, copper mines in the USA, a mine in Bulgaria, coil spoil heaps in England, mine samples in Australia and biooxidation plants in South Africa. *L. ferrooxidans* is as ubiquitous as *T. ferrooxidans* and *T. thiooxidans*.

The important leaching organisms grow in acidic, inorganic habitats and in many instances cannot tolerate more than traces of organic matter.❸ These bacteria could therefore be expected to have evolved in relative isolation from other bacteria. Prior to the advent of DNA and RNA sequencing techniques it was not possible to determine the evolutionary relationship of the acidophilic autotrophs to the rest of the bacterial kingdom. A considerable amount of molecular sequence information has become available within the past 10 years including partial and complete 5S rRNA and 16S rRNA sequences. Based on these sequences *T. ferrooxidans* and *T. thiooxidans* isolates are closely related and have been placed within the proteobacteria at a point close to the division between the β and γ subgroups.

The sequences of a number of other molecules such as the genes for the RecA protein, glutamine synthesizes and the β subunit of the F_1F_0 ATP synthase have been determined from a large number of bacteria. Phylogeny based on the amino acid sequences of the RecA protein, glutamine synthetase and the β subunit of the F_1F_0 ATP synthase have confirmed the classification of *T. ferrooxidans* as a β proteobacterium. An interesting exception is the product of the *nifH* gene. The *T. ferrooxidans* nitrogenase iron protein is clearly most closely related to the equivalent proteins of the genus *Bradyrhizobium*, which is an x proteobacterium, and may be an example of lateral gene transfer.

Lane et al., analyzed partial 16S rRNA sequences of three isolates of *Leptospirillum*-like bacteria. They reported that although the three isolates were closely related to each other (ca. 94% similar), they were not specifically related to any of the existing divisions of bacteria and suggested that the leptospspirilli may represent a new phylogenetic division. Near complete 16S rRNA sequence data of represent a new phylogenetic division. Near complete 16S rRNA sequence data of in the construction of Fig. 1.3. There is a discrepancy in the relationship of the genus *Leptospirillum* to other bacteria if one compares the information in the Ribosome Database Project (WWW) with that in the National Center for Biotech-

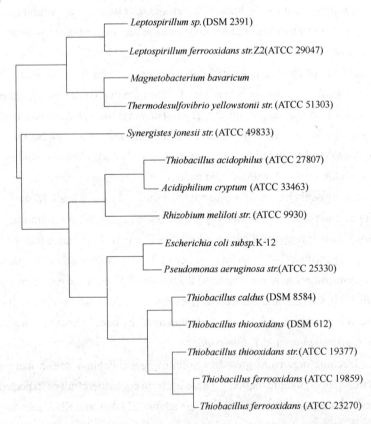

Fig. 1.3 Dendrogram showing the phylogenetic relationship between the relatively closely related bacteria *T. ferrooxidans*, *T. thiooxidans* and *T. caldus* and the distantly related *Leptospirillum* species

(The dendrogram was constructed based on 16S rRNA sequences using the Ribosome Database Project website (http://www.rdp.life.uiuc.edu))

nology Information (NCBI) taxonomy base (WWW). In the NCBI data base the leptospirilli have been placed within the group *Nitrospira* whereas the managers of the Ribosome Database Project have not recognized a *Nitrospira* group. It is interesting that one of the closest bacterial relatives to members of the genus *Leptospirillum* that has been reported so far is the magnetotactic bacterium *Magnetobacterium bavaricum*.

1.4 Nutrition

Bacteria in bioleaching have very modest nutritional requirements. Aeration of a sample of iron pyrite in acidified water is sufficient to support the growth of *T. ferrooxicians* and *L. ferrooxidans*.❹ Air provides the carbon, nitrogen and oxygen source, pyrite the energy source and trace elements, and acidified water the growth environment. *T. thiooxidans* is not able to oxidize ferrous iron to produce the ferric iron required to attack the mineral, however, it readily grows on pyrite in combination with either *T. ferrooxidans* or *L. ferrooxidans*.

1.4.1 Carbon nutrition sources

T. ferrooxidans strains that have been confirmed as being pure are obligate autotrophs. Some early studies showed that after a period of adaptation *T. ferrooxidans* was able to grow on organic substrates and that this was followed by a permanent loss of the ability to oxidize iron. However, the G+C mol% ratio of the cultures changed under these conditions and heterotrophic growth was almost certainly due to the inability of researchers to free their cultures from the presence of the closely associated heterotrophic bacteria belonging to the genus *Acidiphilium*. The iron-dependent, mixotrophic growth of one strain has been reported but unfortunately that isolate has been lost.

Carbon dioxide fixation in *T. ferrooxidans* takes place via the Calvin reductive pentose phosphate cycle. One of the most important enzymes in this process, ribulose 1, 5-biphosphate carboxylase (RuBP-Case) has been characterized. This work also showed that growth on ferrous iron was reduced unless the concentration of CO_2 in the air was increased. This observation is in contrast to the work of others in which it was found that the concentration of CO_2 in air was sufficient to avoid limitation on growth on ferrous iron and mineral sulfide oxidation by *T. ferrooxidans*. The bacterium responds to CO_2 limitation by increasing the cellular concentration of RuBPCase. Indeed, *T. ferrooxidans* strain Fe_1 has two sets of the structural genes for RuBPCase. The two sets are separated by more than 5kb and the nucleotide sequence of the coding region of each set is identical although the flanking regions varied substantially. The RuBPCase gene regulator, RbcR, has been isolated and sequenced. Very little work has been carried out on the enzymology or genetics of CO_2 fixation by either *L. ferrooxidans* or *T. thiooxidans*.

1.4.2 Nitrogen nutrition sources

The study of the nitrogen requirements of bioleaching organisms is complied by the phenomenon that ammonia is highly soluble in acid solutions. Atmospheric ammonia readily dissolves in leach solutions and may provide most, if not all, of the nitrogen required for growth. As little as 0.2mmol/L ammonium has been reported to be sufficient to satisfy the nitrogen requirement of *T. ferrooxidans*. This value will be dependent on the amount of ferrous iron or mineral present in the medium or leach liquor.

High concentrations of inorganic or organic nitrogen are inhibitory to iron oxidation. *T. ferrooxidans* is diazotrophi and is able to reduce atmospheric nitrogen to ammonia. This property was first reported by Mackintosh who demonstrated that *T. ferrooxidaes* was able to incorporate N_2 label into cellular material. It has since been shown that all 15 isolates of *T. ferrooxidans* tested contain the structural genes (*nifHDK*) for the nitrogen fixing enzyme, nitrogenase. The ability to fix nitrogen is therefore almost certainly a general property of *T. ferrooxidans*. The *nifHDK* genes from *T. ferrooxidans* ATCC 33020 have been cloned and sequenced.

There is evidence that *L. ferrooxidans* is also capable of fixing atmospheric nitrogen. Genomic DNA from the *L. ferrooxidans* type strain was reported to give a positive hybridization signal with a *nifHDK* gene probe from *Klebsiella pneumoniae*. *L. ferrooxidans* was also shown to reduce acetylene to ethylene and oxidize ferrous iron to ferric iron at low oxygen concentrations. This ability was repressed by added ammonium ions, which is indicative of the ability to fix nitrogen.

The ability of *T. thiooxidans* to fix nitrogen is uncertain. No hybridization signal was obtained when a *nifHDK* gene probe from *Klebsiella pnuemoniae* was used against chromosomal DNA from *T. thiooxidans* ATCC 8085, but a positive signal was obtained when a *T. ferrooxidans nifHDK* probe was hybridized to an unidentified *T. thiooxidans* isolate.

The role of nitrogen fixation in bioleaching operations is difficult to predict. The dissolution of atmospheric ammonia in acid solutions could provide sufficient ammonium to suppress nitrogen fixation. Furthermore, nitrogen fixation is inhibited under fully aerobic conditions therefore might not occur in a well-aerated leaching operation. In the highly-aerated, high oxidation rate, BIOX® tanks used to pretreat gold-bearing arsenopyrite ores, addition of a small amount of ammonia in the form of low-grade fertilizer is required to enhance mineral oxidation.

1.5 Energy Sources

1.5.1 Iron oxidation

As stated earlier, the energy requirements for growth of both *T. ferrooxidans* and *L. ferrooxidans* are able to be met by the oxidation of ferrous to ferric iron under aerobic conditions.[5] Work by Blake and colleagues on the components of iron oxidation in acidophilic bacteria has revealed that the ability to oxidize iron appears to have evolved several times. At least four unique iron-oxidation mechanisms exist. Two of these mechanisms are found in the mesophilic acidophiles. The pathway for iron oxidation in *T. ferrooxtdans* is characterized by the presence of large amounts of the small copper protein, rusticyanin and c-type cytochromes. Rusticyanin is not detectable in *L. ferrooxidans* or in any of the moderately or extremely thermophilic iron-oxidizers. A novel red cytochrome (cytochrome 579) which is clearly different from cytochrome a-, b-or c-type hemes and not found in the other iron-oxidizers, dominates the electron transport chain of *L. ferrooxidans*. This unique cytochrome was redox active with ferrous sulfate.

The components of the iron-oxidation pathway in *T. ferrooxidans* have been relatively well studied. These are a 92 kDa membrane porin, an Fe(II) oxidase, cytochrome C55v rusticyanin and a cytochrome c oxidase of the aa_3-type. All the above components have been isolated and characterized, the amino acid sequence for rusticyanin has been determined and gene for the Fe (II) oxidase have been cloned and sequenced. The exact order of the components and particularly, the position of rusticyanin in the passage of the electrons is uncertain. In a recent review it has been postulated that the role of rusticyanin is to broaden the electron pathway from cytochrome C552 to the cytochrome oxidase as illustrated below.

$$Fe^{2+} \rightarrow Fe(II) \text{ oxidase} \rightarrow \overset{\longrightarrow \text{ rusticyanin} \longrightarrow}{\text{cytochrome } C_{552}} \rightarrow \text{cytochrome c oxidase} \rightarrow O_2 \quad (1.1)$$

Oxygen electrode measurements have been used to determine the apparent Michaelis constant (K_m) of iron-oxidizers for ferrous iron. A K_m value of 0.25mmol/L for *L. ferrooxidans* was considerably lower than the 1.34mmol/L obtained for *T. ferrooxidans* cells. *L. ferrooxidans* is therefore able to grow more efficiently in an environment with a low concentration of iron. Even more important was the finding that ferrous iron oxidation by *L. ferrooxidans* was much less sensitive to endproduct inhibition by ferric iron (K_i 38mmol/L) than *T. ferrooxidans* (K_i 2.5mmol/L). This implies that in a continuous-flow leaching process, where the quantity of ferric iron in solution is high, *L. ferrooxidans* will have a distinct selective advantage over *T. ferrooxidans*. Indeed, *L. ferrooxidans* has been reported to displace *T. ferrooxidans* when mixed cultures were grown in chemostat cultures on either ferrous sulfate- or pyrite-based media.

1.5.2 Sulfur oxidation

Considerably more energy is available during the oxidation of reduced sulfur compounds when compared with ferrous iron. For example, during the complete oxidation of pyrite (FeS_2), 1 electron is derived from the ferrous iron and 14 electrons from the sulfur moiety. Although the oxidation of the iron component of a mineral is probably the most important aspect of metal solubilization, the oxidation of at least some of the sulfur component occurs. Evidence for this is that the growth yield of *T. ferrooxidans* cells expressed as biomass produced per mole of electrons is considerably higher on pyrite than it is on ferrous iron.

Attempts to investigate the pathways involved in sulfur oxidation by acidophilic bacteria have proved difficult. This is partly because of the chemical reactivity and hence lack of stability of many of the sulfur intermediates. What is clear is that reduced sulfur compounds are oxidized to sulfate and that this results in a decrease in pH. Several enzymes involved in sulfur oxidation have been isolated and this research has been reviewed. Although the nature of some of the reactions such as the conversion of elemental sulfur to sulfite and the nature of many of the intermediate sulfur compounds is unknown, the steps in sulfur oxidation appear to be as illustrated below.

$$S_3O_6^{2-} \rightarrow S_2O_3^{2-} \rightarrow S_4O_6^{2-} \rightarrow S_8 \rightarrow SO_3^{2-} \rightarrow SO_4^{2-} \quad (1.2)$$
$$\uparrow$$
$$S^{2-}$$

Sugio and coworkers have identified a sulfite: ferric iron oxidoreductase in *T. ferrooxidans* and a hydrogen sulfide: ferric ion oxidoreductase (SFORase) in a number of strains identified as *T. ferrooxidans* and *L. ferrooxidans*. *T. ferrooxidans* and *L. ferrooxidans* also possess a sulfide-binding protein. The presence of the hydrogen sulfide: ferric ion oxidoreductase and sulfide-binding proteins in *L. ferrooxi-*

dans was surprising as this organism is incapable of growth in sulfur-based media. Even more surprising was the observation that three different washed *L. ferrooxidans* cell preparations had sulfur-oxidizing activity with a pH optimum profile similar to other sulfur-oxidizing bacteria. These workers confirmed the results of previous researchers that *L. ferrooxidans* was not capable of growth in sulfur-salts medium. The explanation of this paradox is still unknown.

A model has been proposed in which *T. ferrooxidans* is able to reduce elemental sulfur to sulfide using glutathione. This sulfide can be oxidized using Fe^{3+} as an electron acceptor to produce sulfite and Fe^{2+}. An NADH-dependent sulfite reductase has been isolated from the periplasm of iron-grown *T. ferrooxidans* cells which reduces sulfite to hydrogen sulfide and $NADH^+$. The possible function of sulfide cycling in *T. ferrooxidans* is uncertain, but it has been suggested that sulfide acts as an energy storage compound.

1.5.3 Electron donors and electron acceptors

Although *T. ferrooxidans* is considered to be strictly autotrophic, it is capable of growth on formic acid. Small organic acids are toxic to acidophiles because at low pH values the organic acids are present in their undissociated form and in this form organic acids readily diffuse across the cytoplasmic membrane. Once inside the cell, organic acids dissociate because of the near neutral intercellular pH which leads to a dissipation of the proton gradient. The secret of obtaining growth of *T. ferrooxidans* on formate was to ensure that the formate was provided at low concentration in a chemostat. Using this approach, higher cell densities were obtained using formate as an energy source than when using ferrous iron or reduced sulfur compounds. Utilization of formate was via the Calvin cycle and six out of seven *T. ferrooxidans* strains tested were able to oxidize formate with strain ATCC 21834 being particularly active. None of the two *T. thiooxidans* strains tested were able to grow on formate.

T. ferrooxidans has been shown to possess an inducible hydrogenase which enabled the bacterium to grow aerobically using hydrogen as an energy source. The bacterium was less acidophilic when growing on hydrogen (pH range for growth 2.5-6.0) and had a slightly longer doubling time (5.0h vs. 4.5h) than when grown with sulfur or ferrous iron. All three *T. ferrooxidans* strains but not the single *T. thiooxidans* strain tested had the ability to utilize hydrogen.

Metal ions which exist in more than one oxidation state and have redox potentials less than the O_2/H_2O couple, have the theoretical potential to serve as electron donors for the growth of bioleaching bacteria.[6] It has been reported that *T. ferrooxidans* is able to directly oxidize U^{4+} to U^{6+} under aerobic conditions and that this oxidation reaction is coupled to carbon dioxide fixation. Similarly, the direct oxidation of Cu^+ to Cu^{2+} has been coupled to CO_2 fixation. However, whenever iron is present, it is difficult to unequivocally demonstrate the direct oxidation of the metal as opposed to the oxidation of ferrous iron to ferric which may then oxidize the metal chemically. *T. ferrooxidans* has been reported to oxidize Mo^{5+} to Mo^{6+} and the evidence for this is strong as a molybdenum oxidase was purified from cell extracts. Bioleaching bacteria are able to oxidize arsenopyrite ores and the potential exits for the oxidation of As^{3+} to As^{5+} to serve as an alternate electron donor. However, this property has not been conclusively demonstrated. *T. ferrooxidans* may be more versatile with respect to its ability to oxidize metals than is generally recognized. A similar metal-oxidizing capacity has not been demonstrated for either *L. ferrooxidans* or *T. thiooxidans*, but much less research has been carried out with these bacteria.

T. ferrooxidans is also versatile with respect to the compounds and it is able to use electron acceptor. Typically, *T. ferrooxidans*, *T. thiooxidans* and *L. ferrooxidans* all respire using oxygen as a terminal electron acceptor. However, in the absence of oxygen, *T. ferrooxidans* is able to use ferric iron as an electron acceptor provided that a reduced form of sulfur, such as elemental, serves as the electron donor. This means that *T. ferrooxidans* is capable of growth in an anaerobic environment. The practical implication of this is that metal solubilizing activity may take place at the center of poorly aerated dump or heap using the ferric that was produced by at the center of poorly aerated dump or heap using the ferric that was produced by this ferric iron-reducing ability and are therefore probably obligately aerobic. Since the ferrous/ferric iron couple can not be used as both electron donor and acceptor, *L. ferrooxidans* is also likely to be an obligate aerobe.

Besides its ability to reduce ferric iron, *T. ferrooxidans* is also able to reduce donor. The ability to reduce Mo^{6+}, Cu^{2+} and Co^{2+} is a property of the hydrogen sulfide: ferric iron oxidoreductase (SFORase), the same enzyme which couples the oxidation of elemental sulfur to the reduction of Fe^{3+}. Interestingly, a SFORase has also been found in a strain of *L. ferrooxidans* but its role in this bacterium is uncertain.

1.5.4 Mineral oxidating ability

Since *T. ferrooxidans* is able to oxidize both iron and sulfur, it is not surprising that the bacterium is able to solubilize a wide variety of metals from ores in pure culture. Even though leptospirilli are able to use only iron as an energy source, these bacteria are nevertheless able to degrade pyrite very efficiently. This is because both types of organism are able to oxidize ferrous iron to ferric and ferric iron is required for mineral solubilization. At 30°C the growth rate of *L. ferrooxidans* on iron in shaken batch cultures is about half that of *T. ferrooxidans*, however, these two species degrade pyrite at similar rates. In contrast to these bacteria, *T. thiooxidans* is unable to degrade sulfidic minerals in pure culture. However, when *T. thiooxidans* is grown in mixed culture with either *T. ferrooxidans* or *L. ferrooxidans* it may enhance the ability of either pure culture to degrade sulfidic ores.

The relative roles of the mesophilic bacteria in bioleaching operations has been the subject of much research and has been reviewed. More recent work has reassessed the relative roles of the mesophilic bacteria and suggested that *T. ferrooxidans* may not be as dominant as previously thought. In percolation leaching experiments using crushed pyrite or complex sulfidic ores and a temperature of 28°C, *L. ferrooxidans* was at least as numerous as *T. ferrooxidans* and when in pure culture mobilized metals at least as efficiently. Similar conclusions have been reached when molecular DNA analysis techniques have been used to identify bacteria in bioleaching environments. An analysis of the spacer regions between the 16S and 23S rRNA was used to identify the dominant bacteria during the column-leaching of a complex copper inoculated with bacteria from the bottom of an industrial heap-leaching operation at Lo Aguirre, Chile. If ferrous iron was added to the leach liquor, then *T. ferrooxidans* dominated the population. If no ferrous iron was added, the ferrous iron concentration remained low and only *L. ferrooxidans* and *T. thiooxidans* were detected. Under these conditions bioleaching efficiency remained high with

90% of the copper being recovered. In another study, adherence of *T. ferrooxidans*, *L. ferrooxidans* and *T. thiooxidans* to a complex copper sulfidic ore during column-leaching using specific antibodies was measured. *T. ferrooxidans* was dominant for the first 5 days but after 60 days of leaching, *L. ferrooxidans* and *T. thiooxidans* outnumbered *T. ferrooxidans* by 1000-fold.

An investigation of the bacteria present in the tanks of a continuously operating pilot plant treating gold-bearing arsenopyrite ores has revealed similar results. The tanks had been inoculated with leach solution from an industrial plant and were operating at 40°C and pH 1.6. Analysis of the 16S rRNA from these tanks indicated that the population was dominated by *L. ferrooxidans* and *T. thiooxidans* (or *Thiobacillus caldus*) and that if present, the number of *T. ferrooxidans* cells was too low to be detected.

1.6 Mesophilic and Acidophiles in Mineral Bioleaching

It has been argued that at low pH values (pH < 3.5) indirect solubilization of the mineral by ferric iron hexahydrate is the major mechanism of mineral attack. Bacteria such as *T. ferrooxidans* and *L. ferrooxidans* adhere strongly to the surface of a sulfidic ore and are surrounded by an exopolymer layer which is heavily impregnated with irons and polythionate granola (Fig. 1.4). These metal-containing exopolymers give the cells a slightly positive zeta potential and this was necessary for attachment as it overcomes the repulsion between the negatively charged sulfide ore and negatively charged washed cells. The ferric iron hexahydrate molecules at the surface attack the sulfidic ore indirectly (chemically) and ferrous iron and thiosulfate are produced.

The ferrous iron is rapidly reoxidized to ferric by the bacterium and is recycled within the exopolymer layer of the attached bacteria (Fig. 1.4). At low pH the unstable thiosulfate is converted to polythionates and elemental sulfur. Since elemental sulfur is poorly soluble this is the form of sulfur most readily detected, often visually. In the case where *T. ferrooxidans* is the generator of ferric iron, thiosulfate and polythionates may be may be oxidized to sulfate. Where *L. ferrooxidans* is the ferric iron generator, sulfur-oxidizing bacteria such as *T. thiooxidans*, *T. caldus* or even *Thiobacillus acidophilus* and other facultative sulfur oxidizers may be responsible for the oxidation of thiosulfate and sulfur derivatives.

It has been pointed out that the stoichiometry and sulfur intermediate of this equation is different from that in the equation most commonly referred to for pyritic ore attack.

$$FeS_2 + Fe_2(SO_4)_3 \longrightarrow 3FeSO_4 + 2S^0 \qquad (1.3)$$

In general, bioleaching bacteria are remarkably adaptable when faced with adverse growth conditions.[7] There appears to be a substantial variation in tolerance to some growth conditions like pH, temperature, or metal ion tolerance between strains or isolates which have the same genus and species designation. Adapted populations may also differ substantially from the parental cultures. Furthermore, there is some interdependence between variables, e.g., the pH of a medium may affect temperature tolerance.

T. ferrooxidans grows best within the pH range 1.8-2.5. There have been of growth at a pH of 1.5 after selection in continuous culture. *T. thiooxidans* is considerably more resistant to acid and is capable of growth at a pH of less than 0.8. *Leptospirillum* is also more resistant to low pH than *T. ferrooxidans* and will grow at a pH as low as 1.2.

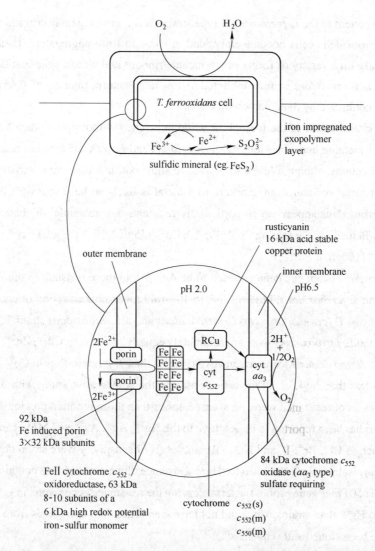

Fig. 1.4 Diagram illustrating ferrous/ferric iron cycling in the exopolysaccharide layer of a *T. ferrooxidans* cell attached to a mineral particle and a model of electron transport path

The optimum temperature for the growth of both *T. ferrooxidans* and *T. thiooxidans* is probably about 30-35°C. However, some strains of *T. ferrooxidans* are adapted to low temperatures. It has been reported that the growth rate halves with each 6°C within the range 25-2°C. Some strains are able to oxidize pyrite at temperatures of as low as 10°C. As may be expected, these cold-tolerant strains are less tolerant of high temperatures than more typical mesophilic isolates. The upper limit for growth of *T. ferrooxidans* is probably close to 40°C. Cultures which grow above 42°C are almost always dominated by mixtures of *T. caldus* and *L. ferrooxidans* rather than *T. ferrooxidans*.

In general, *L. ferrooxidans* strains appear to be more tolerant of high temperatures and less tolerant of low temperatures than *T. ferrooxidans*. *Leptospirillum*-like bacteria have been reported to have an upper limit of about 45°C and a lower limit of about 20°C. A pilot scale (1m³) BIOX® plant operating at constant temperature of 45°C was found to be dominated by bacteria with an identical 16S rRNA restriction enzyme pattern to the *L. ferrooxidans* type strain. When growing on iron at lower temperatures (15-20°C), *L. ferrooxidans* cells become embedded in slime to form aggregates. These macroscopic aggregates can take on a variety of forms which include ribbons and almost spherical-like pellets. The observation that *L. ferrooxidans* is more inhibited by low temperatures than either *T. ferrooxidans* or *T. thiooxidans* was confirmed by Sand and coworkers.

An important characteristic of the mesophilic acidophilic chemolithotrophs is their tolerance of high concentrations of metallic and other ions. *T. ferrooxidans* and *L. ferrooxidans* are resistant to a large number of metal cations, although levels of resistance show considerable strain variation. Adaptation to high levels of metal resistance on exposure to a metal is likely to be responsible for much of the variation. *T. ferrooxidans* appears to be particularly resistant. For example, the bacterium has been reported to grow in medium containing Co^{2+} (30g/L), Cu^{2+} (55g/L), Ni^{2+} (72g/L), Zn^{2+} (120g/L), U_3O_8 (12g/L) and Fe^{2+} (160g/L).

One *L. ferrooxidans* strain has been reported to be 4-to 5-fold more resistant to uranium, silver and molybdenum than *T. ferrooxidans*, but sensitive to 100-fold lower concentration of copper. In a comparative study of two *T. ferrooxidans*, two *L. ferrooxidans* and a *T. thiooxidans* strain, it was found that *T. ferrooxidans* and *L. ferrooxidans* were approximately equally resistant to Cu^{2+}, Zn^{2+}, Al^{3+}, Ni^{2+} and Mn^{2+}, but that *L. ferrooxidans* was more sensitive (<2g/L) than *T. ferrooxidans* to Co^{2+}. *T. thiooxidans* was sensitive to less than 5g/L of all the cations used in the comparative study with the exception of Zn^{2+} (10g/L). Levels of resistance were, on average, lower than those reported previously.

T. ferrooxidans has been reported to be sensitive to the Hg^{2+}, Ag^+, As^{3+} and Mo^{6+} cations and monovalent anions such as Cl^-, Br^-, I^- and NO_3^-. Again, levels of sensitivity were strain dependent. When ten *T. ferrooxidans* isolates were screened for Hg^{2+} resistance, three of the strains contained DNA which hybridized to a Tn501 *mer* gene probe. Bacteria carrying the resistance genes were, in general, 3-5 times more resistant to Hg^{2+} than strains which did not have *mer* genes. The *mer* genes from *T. ferrooxidans* strain E-15 have been cloned and sequenced.

Although normally sensitive to As^{3+} and to a lesser extent As^{5+}, bioleaching bacteria can be adapted to high concentrations of As ions. For example the BIOX® process for the treatment of refractory gold-bearing arsenopyrite ores operates with a total As concentration of >13g/L. Adaptation to as was achieved by exposing the bacteria to increasing concentrations of arsenic in continuous flow reactors. Recently, the As genes from *T. caldus* have been identified and the As resistance genes from *T. ferrooxidans* ATCC 33020 have been cloned.

Sensitivity of bioleaching organisms to chloride ions is of considerable economic importance. For example, in the dry parts of western Australia or northern Chile most of the available ground water is high in chloride and the use of this water in leaching process make-up water is therefore limited. A mixture of thiobacilli has been used to leach a sulfidic copper ore in the presence of 5g/L Cl^- but attempts to adapt *T. ferrooxidans* to > 5g/L Cl^- have not been successful.

1.7 Summary

The mesophilic iron- and sulfur-oxidizing autotrophic bacteria described in this chapter play an essential role in many currently operating commercial biooxidation processes. However, the number of types of bacteria that will be used in future processes is likely to increase substantially as new processes are developed. Bacteria capable of oxidizing ores at higher temperatures are particularly important because at higher temperatures the chemistry of microbially-assisted leaching is much faster. This means that several processes which are uneconomic at lower temperatures are likely to become financially viable. Each new organism will have strengths and weaknesses which are likely to be different from the mesophilic bacteria described in this chapter. Studies on moderately and extremely thermophdic iron- and sulfur-oxidizing organisms are less advanced than those with the mesophdic bacteria, but in the future there is likely to be a range of organisms available, each with different advantages and disadvantages which will need to be considered for use in the biooxidation of a given ore.

Bioleaching and biooxidation have emerged as accepted commercial practices that offer the industry cost-effective, simple, robust, high performance and environmentally friendly alternatives to conventional mineral and metallurgical processing methods. Aerated, stirredtank biooxidation and heap biooxidation/bioleaching are used throughout the world to process refractory gold ores and concentrates and chalcocite ores. Minerals bioremediation processes, also enjoying commercial success, have emerged as alternatives to more expensive and cumbersome chemical/physical processes. New fundamental and applied developments in bioleaching/biooxidation and minerals bioremediation promise to extend technology applications, and reduce costs and enhance performance. In the near future, we can look forward to greater commercial application of mineral bioprocesses, incorporating innovations that will assist the industry in remaining competitive, and processing in an environmentally responsible way.

Questions

1. What is the difference between bioleaching and bio-oxidation?
2. How does CO_2 fixation work as nutrition to bacteria?
3. What are the nutrition sources for bacteria in bioleaching?
4. How does pH value affect bacteria growth?
5. What is the suitable temperature for different bacteria?

Note

注　释

❶ 微生物浸出是借助于微生物的氧化作用把有价金属从矿石浸出到溶液中，然后处理溶液获取最大金属回收率，并丢弃剩余固体残渣。

❷ 矿石的生物氧化是采用同样的浸矿细菌对矿石中的硫化物进行催化降解的预处理过程。

❸ 重要的浸矿生物体一般生长在酸或无机物环境中，在许多情况下都难以与微量有机物共存。

❹ 细菌在浸矿过程中所需营养不多，对酸化溶液中的黄铁矿进行通气即可满足氧化亚铁硫杆菌和氧化亚铁钩端螺菌的生长。

❺ 在通气条件下，氧化亚铁被氧化成三价铁离子的过程即可提供氧化亚铁硫杆菌和氧化亚铁钩端螺菌生长所需的能量。

❻ 多于一个氧化电位的金属离子其氧化还原势低于 O_2/H_2O 时，理论上更易提供电子供浸矿细菌生长。

❼ 通常，浸矿细菌能够有效地适应不利的生长条件。

2 Surface Chemistry of Bacterial Leaching

本章要点　细菌在矿物表面吸附是直接浸出机理实现的先决条件，这一过程是范德华力和静电力共同作用的结果。细菌生长繁殖在矿物表面形成了一个离子浓度、营养物质都与矿物本体不同的微环境，细菌持续的生长会导致矿物表面形成生物质膜。

2.1 Introduction

An understanding of the mechanisms and physical chemistry of bacterial leaching could be profitably used in the design and optimization of bioremediation and metal extraction processes.

Two broad mechanisms of bacterial leaching have been proposed: (1) the "direct" mechanism, in which bacteria attach to the surface of the mineral, and attack the mineral at the point of attachment, and (2) the "indirect" mechanism, in which the role of the bacteria is to oxidize ferrous ions to ferric ions, and the ferric ions dissolve the sulfide mineral.

The direct mechanism of bacterial leaching of mineral sulfide, for example, sphalerite, can be represented by the overall reaction:

$$ZnS + 1/2O_2 + 2H^+ \xrightarrow{\text{bacteria}} Zn^{2+} + S + H_2O \tag{2.1}$$

whereas the indirect mechanism requires the presence of iron in solution:

$$ZnS + 2Fe^{3+} \longrightarrow Zn^{2+} + S + 2Fe^{2+} \tag{2.2}$$

$$4Fe^{2+} + 4H^+ + O_2 \xrightarrow{\text{bacteria}} 4Fe^{3+} + 2H_2O \tag{2.3}$$

In both cases, bacteria are able to utilize sulfur as a substrate. This reaction is given as:

$$2S + 3O_2 + 2H_2O \xrightarrow{\text{bacteria}} 2H_2SO_4 \tag{2.4}$$

Both mechanisms of bacterial leaching are surface processes. The attachment of bacteria to the mineral surface is affected by hydrophobic and electrostatic parameters and can be treated by the methods of surface chemistry. All dissolution reactions in aqueous solutions include an electrochemical step at the surface, and step often controls the rate of reaction.

The aim of this chapter is to discuss the physical chemistry of bacterial leaching. One topic is examined here in particular: the surface chemistry of bacterial attachment. The experimental results and a theoretical interpretation of the surface phenomena are outlined for each topic.

2.2 Adhesion and Attachment of Bacteria to Mineral Surfaces

The attachment of bacteria to a mineral surface is a prerequisite for the leaching of minerals by the direct mechanism. Indeed, the prime evidence of the direct mechanism is, at this stage, that *Thiobacillus ferrooxidans* attaches to mineral surfaces.❶ On its own this is unconvincing evidence because bacterial cells generally adhere readily to many types of surfaces. There is little consensus about the significance of bacterial attachment in microbial leaching. For example, Espejo and Ruiz determined that the total activity associated with the direct mechanism to between 1%-10%, while Boogerd et al. determined that 83%-92% of the pyrite was oxidized by the direct mechanism. The clarification of the role of bacterial attachment to mineral surfaces would provide a significant step towards the effective design and operation of bacterial leaching processes.

The colonization of a surface by bacteria is visualized as occurring in four distinct steps. These steps are illustrated in Fig. 2.1. In the first step bacteria are transported to the surface by diffusion, convection or active movement. Once in the vicinity of the surface, the next step is the initial attachment or adhesion of the bacteria to the surface. This process is thought to be dominated by physicochemical forces, such as the electrostatic and hydrophobic forces between particles and surfaces. The terms initial attachment and adhesion are used interchangeably to describe this second step. The attachment of bacteria to the surface is the third step in the process. Bacteria that adhere to the surface may develop special structures such as fibrils, or secrete polymers such as polysaccharides to become more firmly attached to the surface. The fourth and final process is the colonization of the surface by the formation of microcolonies and biofilms.

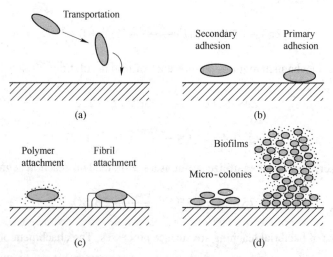

Fig. 2.1 The four stages in the colonization of a surface by bacteria
(a) transport of bacteria to the surface; (b) adhesion or initial attachment of bacteria to the surface at either the primary or the secondary minima; (c) attachment of bacteria to the surface by the excretion of lipopolysaccharides or by the growth of fibrils; (d) the formation of micro-colonies and biofilms of bacteria

The study of bacterial attachment to and biofilm formation on minerals is a developing field and the methods of surface and colloid science are just beginning to throw new light on the processes of bacterial leaching. Physicochemical aspects of the adhesion of bacteria to the surface of minerals, the formation of biofilms on pyrite and the formation of colloids during bacterial leaching are discussed below.

The attachment of bacteria to a mineral can be divided into two steps: (1) the initial attachment, which is governed by surface properties of the mineral and the bacteria; and (2) the specific attachment, which is the growth of appendages or proteins that anchor the bacterium to the surface. The initial attachment can be reversible or irreversible. Reversible attachment is the weaker of the two, and the bacterium can move back into the solution. A dynamic equilibrium exists between the reversibly adsorbed bacteria on the surface and those in the solution, and most mathematical models of bacterial leaching include terms describing this equilibrium. A cell that is irreversibly attached does not move along the surface nor can it be removed by moderate shear forces.

Van Loosdrecht et al. have shown that hydrophobic and electrostatic forces determine the initial attachment of many different types of bacteria to surfaces. They examined the adhesion of more than 20 types and strains of bacteria to polystyrene and found that the surface coverage by the bacteria was more strongly correlated with the contact angle of the bacteria than with the electrophoretic mobility. This suggested that the hydrophobic interaction was more important than the electrostatic force.

Van Loosdrecht and Zehnder summarized the previously published results as follows: (1) adhesion increases with increasing hydrophobicity of the bacterium or solid; (2) adhesion decreases with increasing electrostatic repulsion; (3) usually hydrophobicity is more important than electrostatic interactions; (4) adhesion is generally reversible; (5) irreversible adhesion occurs when the electrostatic interaction is weak or when the hydrophobicity is strong; and (6) there is evidence of a layer of water between the bacterium and the surface.

The mechanism of adhesion of *T. ferrooxidans* to minerals has not been thoroughly investigated. Blake et al. found that the cells grown at pH 2.0 on pyrite or ferrous ions were negatively charged, while those grown on sulfur were close to their isoelectric point. Both pyrite and sulfur are negatively charged at pH 2.0. In spite of the electrostatic repulsion, mixtures of cells and pyrite or cells and sulfur aggregated to form a colloid with properties of an intermediate of the two starting materials. Blake et al. did not examine the hydrophobic interactions. Solari et al. found that *T. ferrooxidans* is slightly hydrophobic, and that the contact angle increases when the pH decreases from 3.0 to 2.0. They showed that the bacteria adhere preferentially to hydrophobic surfaces such as mineral sulfides. This adhesion is weak, representing reversible adsorption of the bacteria to the mineral surface.

The second step in the process of bacterial attachment is the development of a means of specific attachment to minerals. Arrendondo et al. removed part of the polysaccharide material that surrounds the *T. ferrooxidans* cell wall and showed that this resulted in an increased number of cells that adhere to sulfide minerals. They concluded that this was not only due to the increased hydrophobicity of the cell wall, but that proteins in the cell wall play an important part in the mechanism of adhesion. This view is supported by Devasia et al. who concluded that an appendage made of protein is present on the surface of the cell and is responsible for adhesion of *T. ferrooxidans* to sulfide minerals. The structures or proteins that would allow for such specific interactions in *T. ferrooxidans* are yet to be identified.

2.3 Theory of Bacterial Attachment to Surfaces

Bacteria in solution have been frequently described as colloidal suspensions and the adhesion of bacteria has been treated in terms of the theory of colloid chemistry. The physical chemistry of colloids treats the long-range interactions of a colloidal particle and a surface as the combination of the van der Waals force and the electrostatic force.

The van der Waals force arises when the oscillating charges of the positive nucleus and the negative electrons in an atom (or molecule) cause an instantaneous dipole moment and induce an instantaneous dipole moment in another atom (or molecule). This force is also responsible for the long-range attractive forces, between macro-bodies, such as that between a colloid and a surface. The presence of a medium such as water alters the magnitude, but not the dependence on distance of the van der Waals force. Hydrophobicity arises when the magnitude of the van der Waals interactions between water molecules is greater than the interactions of the water molecules with the colloid and with the surface. The hydrophobic interaction is effectively the solvent expelling the colloid to the surface.

The interaction energy, V_A, arising from the van der Waals force between a spherical colloid particle of radius, r, separated from a planar surface by a distance, H, can be expressed as:

$$V_A = -\frac{A_{132}r}{6H}\left[1 + \frac{H}{H+2r} + \frac{H}{r}\ln\left(\frac{H+2r}{H}\right)\right] \tag{2.5}$$

where A_{132} is the Hamaker constant of the system. The Hamaker constant is dependent on the polarizability and the number density of the atoms of each of the components of the system. The Hamaker constant of the system is given as:

$$A_{132} = A_{12} + A_{33} - A_{13} - A_{23} \tag{2.6}$$

where A_{12} is the Hamaker constant for the surface-colloid interaction, A_{33} for the water-water interaction, A_{13} for the surface-water interaction, and A_{23} for the colloid-water interaction. Thus the presence of the water will increase the colloid surface interaction if A_{132} is larger than A_{12}, that is, if A_{33} is larger than $A_{13} + A_{23}$. In other words, if the water-water interactions are large, then the presence of water increases the Hamaker constant, and the effect is that the colloid is expelled from the water towards the surface. This is the hydrophobic interaction.

The hydrophobicity of the bacterial cell wall has been determined using a number of techniques, the most popular of which are the measurement of the contact angle of a drop of liquid on a layer of cells, the liquid-liquid partitioning of cells in an aqueous-hydrocarbon two-phase system and hydrophobic interaction chromatography. The measurement of the contact angle has been the most common method of evaluating the hydrophobicity of the bacterial cell in studies related to bacterial leaching. The contact angle is measured directly using a goniometer. Cells are placed on a substrate, such as a glass slide or a filter membrane, then a drop of solution is placed on top of the layer of cells. The angle between the meniscus of a drop of solution and a layer of bacterial cells is then measured.

However, the relationship between the hydrophobic interaction and measurable quantities such as the contact angle is not straightforward. The Hamaker constants can be calculated from the contact angle by a method outlined by Fowkes. He described a method for calculating the contribution of the dispersion forces to the surface free-energy of a substance from the contact angle and showed that this quantity can be used to calculate the Hamaker constants. In contrast, van Oss et al. obtained an expression for the contact angle and the Hamaker constant in terms of the Lifshitz-van der Waals component of the surface free-energy.

The electrostatic force, which is generally repulsive in bacteria-surface interactions, arises because both the surface and the colloid are charged. The development of a surface charge (and as a result, a potential difference between a solid and the solution) is one of the principle properties of surfaces. The surface charge originates from four different phenomena: (1) the termination of bonds of the solid lattice at the surface may result in uncompensated charge; (2) the degree of ionization of function groups such as -SH at the surface and thus the surface potential may be dependent on the pH of the solution; (3) the polarization of the surface as a result of charge-transfer reactions; and (4) the adsorption of ions or hydrophobic species on the surface.

The solid-solution phase boundary consists of a double-layer structure in which the charge on the solid surface is balanced by the charge in the solution. Three double layers exist at the solid-solution boundary. These are the space-charge layer which forms on the mineral side of the boundary, the compact Stern layer which forms between the solid surface and the distance of closest approach of nonadsorbed ions, and the diffuse Gouy layer in the solution. Fig. 2.2 illustrates the formation of the compact Stern layer and the diffuse Gouy layer. The compact Stern layer is represented by the layer between the surface and the outer Helmholtz plane, which is assumed to be sufficiently close to the plane of shear for them to be regarded as occurring at the same position.

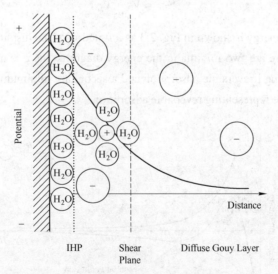

Fig. 2.2 Schematic representation of a possible distribution of potential at an interface
(IHP indicates the inner Helmholtz plane, located at the surface of a layer of adsorbed ions. Anions are shown here as unhydrated, and cations as hydrated. The plane of shear is beyond, but sufficiently close to the next layer of adsorbed ions (referred to as the outer Helmholtz plane) for the two to be regarded as occurring at the same position)

The interaction energy, V_E, arising from the electrostatic force between a spherical colloid of radius r separated from a flat surface by a distance H is given as:

$$V_E = \pi\varepsilon\varepsilon_0 r\{(\psi_{13} + \psi_{23})^2 \ln[1 + \exp(-\kappa H)] + (\psi_{13} - \psi_{23})^2 \ln[1 - \exp(-\kappa H)]\} \quad (2.7)$$

where is the potential difference between the surface (1) and the solution (3), ψ_{23} is the potential difference between the colloid (2) and the solution, κ is the inverse of the thickness of the diffuse Golly layer, ε is the dielectric constant and ε_0 is the permittivity of free space.

The potential difference of the double layer cannot be directly measured, instead different techniques are used in different fields of study. The potential difference can be determined with respect to a reference electrode, such as the hydrogen electrode or calomel electrode, or it can be estimated from the surface charge by calculation using the Golly-Chapman theory of the diffuse double-layer.[3]

The measurement of the zeta potential is often used in surface and colloid sciences to determine the surface potential. The zeta potential represents the potential between the plane of shear and the bulk solution. Mineral particles are inert in these studies and the surface potential, and hence the zeta potential, is affected by the adsorption of ions including H^+ and OH^-, and surfactants. Typical measurements show the effect of an adsorbed species, such as a surfactant, as a function of pH. However, it is difficult to determine accurately the distance between the plane of shear and the surface of the particle, and as a result it is difficult to determine the surface potential accurately. Consequently, the surface potentials in Eq. 2.7 are often replaced by the values of the zeta potential. The zeta potential and the electrophoretic mobility are related to each other by the Smoluchowski equation.

The total interaction energy, V_{Tot}, is the sum of the van der Waals energy and the electrostatic energy, and can be given as:

$$V_{Tot} = V_A + V_E \quad (2.8)$$

The total interaction energy is shown in Fig. 2.3 as a function of the distance of separation, H. This figure indicates that there are two minima in the energy diagram: there is a primary minimum close to the surface representing irreversible adsorption and a secondary minimum at a distance of between 5-20 nm from the surface representing reversible adsorption.

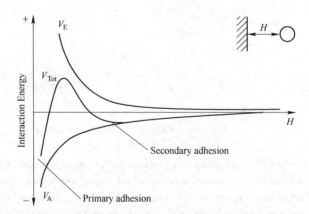

Fig. 2.3 Interaction energy for the long range forces between a spherical colloidal particle and a planar surface as a function of the distance of separation
(The curve of total interaction energy indicates the possibility of two minima at different distances from the surface)

Solari et al. compared the contribution of the hydrophobic interaction and the electrostatic interaction to the adhesion of *T. ferrooxidans* on various mineral surfaces. They determined the hydrophobicity of the bacteria by measuring the contact angle of a layer of cells deposited on a surface and by liquid-liquid partitioning. The electrostatic interaction was determined by the measurement of the electrophoretic mobility and it was reported that the water contact angle of the bacterial cell wall was 20.3° on a fresh layer of cells and 23.3° on a dry layer of cells. Between 14%-22% of the cells were extracted into hexadecane during liquid-liquid partitioning tests. These results were interpreted as indicating that the bacterial cell wall is slightly hydrophobic. Of the minerals tested by liquid-liquid partitioning, only quartz was considered hydrophilic since all of the quartz remained in the aqueous phase. Chalcopyrite and pyrite were concentrated at the interface between the aqueous and the hydrocarbon phases, indicating that these minerals are hydrophobic even though their contact angles are about 60°. The isoelectric point of the bacterial cells occurred at a pH of about 3.0 and that of most of the sulfide minerals between pH 2.0-3.0. Therefore both the sulfide mineral and the bacterium have the same charge at pH 2.0 and below and the electrostatic force is repulsive.

Solari et al. determined that the bacterial adhesion is reversible and corresponds to adsorption at the secondary minimum. They concluded that the hydrophobic interactions dominate the adsorption of *T. ferrooxidans* to mineral surfaces. In addition to these measurements, Solari et al. also determined the kinetics and equilibrium of bacterial adhesion to various minerals. They showed that bacterial adhesion was complete after about 4-6h, which was relatively slow compared to work of others. The isotherm of the adhesion of *T. ferrooxidans* to quartz was close to linear. This isotherm indicated that about 30% of the bacteria adhere to the quartz irrespective of the concentration that is initially present. The coverage of the surface of the minerals by *T. ferrooxidans* was 23.0%, 20.0% and 9.8% on chalcopyrite, pyrite and quartz, respectively. It was concluded that the bacteria attached preferentially to the hydrophobic sulfide minerals rather than to the hydrophilic quartz.

Ohmura et al. compared the attachment of *T. ferrooxidans* and *Escherichia coli* to different minerals. The contact angles of *T. ferrooxidans* and *E. coli* were approximately 23° and 31°, respectively, and these values agreed with those reported previously. However, while 60% of the *E. coli* were extracted in hexadecane in aqueous-hydrocarbon partition tests, only 25% of the *T. ferrooxidans* were extracted in the hexadecane. Both the measurements of the contact angle and the liquid-liquid partitioning indicated that *T. ferrooxidans* cells were more hydrophilic than *E. coli* cells. Ohmura et al. also measured the contact angles of minerals tested. The values of the contact angle with pH 2.0 solution were 28.4°, 68.9°, 83.4° and 80.9° for quartz, pyrite, chalcopyrite and galena, respectively. The adhesion studies indicated that 9.9%, 18.9%, 26.8% and 30.8% of the *E. coli* cells adhered to quartz, pyrite, chalcopyrite and galena, respectively. These results suggested a relationship between the contact angle of the bacterial cells and the contact angle of the mineral. However, results for the adhesion of *T. ferrooxidans* onto these minerals were more variable. Ohmura et al. determined that 4.7%, 24.0%, 14.0% and 1.4% of the *T. ferrooxidans* cells adhered to quartz, pyrite, chalcopyrite and galena, respectively. They concluded that there was a special interaction between *T. ferrooxidans* and the reduced Iron in the minerals, that is *T. ferrooxidans* recognized the reduced iron in pyrite and chalcopyrite and adhered selectively to these minerals but not to galena, by a strong interaction. They believed that this interaction was not physicochemical in nature.

In order to test their hypothesis Ohmura et al. added ferrous ions to the solution which reduced the number of cells that adhered to pyrite. Superficially this confirmed their reasoning, however this result could be explained as follows. The zeta potential for both pyrite and *T. ferrooxidans* is negative under the conditions of their experiments and the addition of iron to the solution would reduce the potential of the mineral surface. Thus, with the addition of ferrous ions to the solution the surface of the mineral would become more negative and the electrostatic repulsion between the cells and the mineral would increase thereby reducing the number of attached cells.

Devasia et al. determined the adhesion of *T. ferrooxidans* that had been grown on different substrates on various minerals. It was observed that about 20% of the cells grown in sulfur, pyrite and chalcopyrite were extracted into the hexadecane phase in liquid-liquid partitioning experiments, while negligible numbers of cells grown on ferrous ions were removed from the aqueous phase. They reported a similar trend for the results from hydrophobic interaction chromatography using octyl-Sepharose as the extractant. About 17% of the cells grown on ferrous ions were bound to the octyl-Sepharose, while about 50% of the cells grown on sulfur, pyrite, and chalcopyrite were bound to the octyl-Sepharose. The isoelectric point of sulfur, pyrite and chalcopyrite moved to higher pH values (approximately from 2.5-4.5) when bacterial cells were present on the surface. It was concluded that both electrostatic and hydrophobic interactions are important in the adhesion of *T. ferrooxidans* to minerals.

Devasia et al. also reported the growth of a surface component when cells were grown on sulfur, pyrite and chalcopyrite. This component was not present when the cells were grown on ferrous ions. This result was obtained by raising antibodies to cells grown on sulfur. Devasia et al. argued that the FTIR difference spectrum indicated that this component was a protein. However, the quantity of lipopolysaccharide on the cell surface is expected to increase with attachment to and growth on a solid substrate such as sulfur and mineral sulfides and Devasia et al. did not compare the spectrum with that of the lipopolysaccharide produced by attached cells.

Sam et al. measured the hydrophobicity of cell suspensions by liquid-liquid partition in aqueous and organic (m-xylene) phases. They measured the electrostatic interactions by determining the zeta potential of the suspensions. They showed that cells grown on sulfur, pyrite, and chalcopyrite exhibit higher hydrophobicity than cells grown on ferrous ions.

From this presentation of work reported in the literature, it is clear that there is agreement on some key parameters. Measurements of the zeta potential of cells of *T. ferrooxidans* and the particles of minerals indicate that their surfaces are negatively charged and that the electrostatic interaction between cells and minerals is therefore repulsive. The measurements of the contact angle of *T. ferrooxidans* consistently yield values of 20-23°C, and the liquid-liquid partitioning with hexadecane as the hydrocarbon results in extraction of 20%-25% of the cells into the organic phase. The adhesion isotherm appears to be linear. However, the surface coverage by the cells is low ranging between 1%-25% and it does not appear to be directly to the contact angle of the mineral.

Sampson et al. investigated the mechanism of the attachment of *Thiobacillus ferrooxidans* and moderately thermophilic bacteria to sulfide minerals. *T. ferrooxidans* (DSM 583 and 4TCC 23270) and four strains of moderate thermophilic bacteria, *Sulfobacillus thermosulfidooxidans*, (strain TH1) and *Sb. acidophilus* (strains THWX, ALV and YTF1) all grown on ferrous iron, sulfur and a chalcopyrite concentrate (termed chalconc) were investigated using 3 sulfide mineral systems: pyrite, a chalcopyrite concentrate (chalconc) and an arsenic containing concentrate (termed arsenoconc). The properties including the growth substrate, mineral species and the cell hydrophobicity were found to have effect on the attachment of the ceils to the mineral surface.

In Roel Cruz et al.'s research, they found that an acid dissolution process can affect the surface characteristics of arsenopyrite and bacterial attachment process. Fig. 2.4 shows a comparison of the bacterial attachment to arsenopyrite and pyrite surfaces which is indirectly expressed by the concentration of planktonic cells in the nutrient medium after inoculation. The number of planktonic cells of *A. ferrooxidans* decreased following inoculation on the pyrite, indicating immediate adhesion to the pyrite surface. A delay in the bacterial attachment was observed for the arsenopyrite pulpthough. For arsenopyrite, the decrease in the number of planktonic bacteria occurred after 70h following inoculation. The study of physicochemical properties of the bacteria/mineral/solution interface at the optimum pH for bacterial growth (1.8-2.5) showed that even though the results of electrostaticand hydrophobic interactions favored bacterial adhesion of *A. ferrooxidans* on arsenopyrite that is not the result observed. A possible explanation for the limited bacterial attachment on arsenopyrite has been given on the basis of an acid dissolution process, that modifies the mineral structure and provokes a preferential dissolution of Fe (II). Once there is Fe (II) in solution, it becomes a more available source of energy for *A. ferroxidans* and generates the Fe (III) that produces the non-contact dissolution of arsenopyrite.

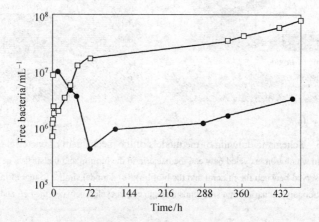

Fig. 2.4 Evolution of planktonic bacteria after inoculation in arsenopyrite (□) and pyrite (●)
(Planktonic cells/mL as a function of time)

There is little consensus about the mechanisms affecting bacterial attachment to minerals. Hydrophobic interactions are clearly important but they are difficult to quantify and no clear pattern has emerged. A significant group of workers believe that the attachment of cells to the mineral involves a specific biochemical interaction rather than a physical interaction and they postulate that components of the cell wall are able to determine the source of energy.

2.4 Biofilm Colloid Formation on Bacterial Cells

Once bacteria adhere to the surface, colonization of the surface begins. Bacteria may develop special cell structures such as fibrils and secrete polysaccharides to attach more securely to the surface. Other responses, such as a loss of flagella may occur. The bacteria grow and form micro-colonies, creating a local micro-environment in which the concentrations of ions and nutrients at the surface are different from those in the bulk. Stable growth on the surface results in the formation of biofilms.

The formation of biofilms of *T. ferrooxidans* has not received much attention. Crundwellexamined the colonization of a polished section of pyrite by a mixed culture of iron-oxidizing bacteria using a confocal microscope. Clusters of microculonies were seen within 3-4 days. After a period of 13 days, the surface was completely covered by a biofilm of 20-50μm thick. A layer of ferric hydroxide formed below the biofilm. Vertical sections of the biofilm showed that the bacteria were concentrated on the solution side of the biofilm.[4] Crundwell bacteria did not require direct attachment to unreacted pyrite in order to utilize it as an energy source. Rather, he speculated that iron was cycled between the bacteria and the pyrite within the biofilm.

Sand et al. proposed a similar model except that iron was cycled in the lipopolysaccharide layer between a single bacterium and the pyrite surface rather than in a colony of bacteria. The models of Sand et al. and Crundwell are complementary and are shown in Fig. 2.5.

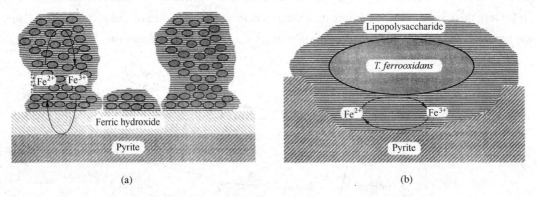

Fig. 2.5 Schematic drawing of the models of the mechanism of bacterial leaching
(a) the biofilm model in which iron is cycled between the bacteria in the biofilm and the surface of the mineral (Corrosion products have been observed between the mineral and the biofilm); (b) a model similar to that of the biofilm model except that the iron in the lipopolysaccharide that surrounds a single cell is cycled between the cell and the mineral surface

The models of bacterial leaching discussed here fall between the classical models of direct and indirect leaching. These newer models suggest that it may be possible to increase the rate of bacterial leaching by increasing the number of bacteria that are attached to the surface and by maintaining a high redox potential in solution.

Two sets of workers have reported the formation of colloids on the surface of *T. ferrooxidans*. Rojas-Chapana et al. detected colloidal sulfur particles of between 4-70 nm in size in the organic capsule around the cell wall. No doubt these sulfur colloids were intermediates in the dissolution process, and were precipitated and stabilized by the lipopolysaccharides surrounding the cell. Rojas-Chapana et al.

speculated that these colloids may serve as an energy reserve. However, Pooley and Shrestha detected colloidal particles of silver sulfide surrounding the bacterial cell wall when silver was added to the leaching solution. The silver sulfide was probably precipitated in the lipopolysaccharide surrounding the cell. However, it is difficult to explain why this silver sulfide was not dissolved by the leaching solution. The detection of colloids of silver sulfide surrounding the bacterial cells seems to suggest that the presence of colloids of sulfur may not be a special feature of the survival strategy of *T. ferrooxidans*; instead they simply form as a result of the reduced solubility of sulfur compounds in the lipopolysaccharide and the cell periplasm.

2.5 Summary

There are two steps of attachment of bacteria to a mineral: (1) the initial attachment, and (2) the specific attachment. The initial attachment can be reversible or irreversible. In reversible attachment the bacterium can move back into the solution. A dynamic equilibrium exists between the reversibly adsorbed bacteria on the surface and those in the solution, and most mathematical models of bacterial leaching include terms describing this equilibrium. A cell that is irreversibly attached does not move along the surface nor can it be removed by moderate shear forces.

The study of bacterial attachment to and biofilm formation on minerals is a developing field and the methods of surface and colloid science are just beginning to throw new light on the processes of bacterial leaching. Physicochemical aspects of the adhesion of bacteria to the surface of minerals, the formation of biofilms on pyrite and the formation of colloids during bacterial leaching are discussed below.

Questions

1. What is the broad mechanism of bacterial leaching?
2. How do the bacteria colonize the mineral surfaces?
3. How is the hydrophobicity of the bacterial cell determined?
4. What is the solid-solution phase boundary like when the solid surface is balanced by the charge in the solution?
5. How do you think about the theory of bacterial attachment to mineral surfaces?

Note

注 释

❶ 细菌在矿物表面吸附是直接浸出机理实现的先决条件，所以在这个阶段氧化亚铁硫杆菌在矿物上的吸附，是其直接浸出机理的主要证据。

❷ 溶液中的细菌可以被认为是一种胶状悬浮体，细菌的吸附过程可以用胶体化学的理论来分析。胶体物理化学认为胶状颗粒在表面的吸附是范德华力和静电力共同作用的结果。

❸ 双层电势差不容易被测量，因而在不同领域内采用不同的测量技术。电势差可以用参比电极进行测量，例如氢电极、甘汞电极或者用双扩散层的 Golly-Chapman 理论计算表面电荷来推算。

❹ 细菌的生长繁殖在矿物表面形成了一个离子浓度、营养物质都与矿物本体不同的微环境。细菌持续的生长会导致矿物表面形成生物膜。

3 Electrochemistry of Mineral Dissolution and Bioleaching

本章要点 为研究方便,可将反应分为阴阳极过程,但这并不是表示两个在矿物不同位置发生的反应。研究黄铁矿的浸出行为对研究细菌浸出机制有重要作用,并且研究黄铁矿溶解的电化学机制也有助于对细菌浸出的理解。通过浸出电化学机制可得出混合电势表达式,矿物溶解电势可以通过参比电极来测量,并采用此方法研究溶液变化对浸出反应式的影响。

3.1 Introduction

Dissolution is the process in which material moves from the solid phase, across the charged phase boundary and into the liquid phase. Chemical bonds are broken in the solid and are formed in solution. The rate of dissolution is affected by the bonding in the solid, the movement of ions and electrons across the phase boundary and the formation of bonds in solution.

Dissolution reactions may occur in the presence or absence of a redox couple. Those dissolution reactions that occur in the absence of a redox couple involve only the transfer of ions. There is no change in oxidation to state and they are referred to as nonoxidative reactions. Dissolution reactions in the presence of a redox couple involve the transfer of ions and are referred to as oxidative reactions when the mineral is oxidized and as reductive reactions when the mineral is reduced. All of these types of reactions are common in leaching chemistry.

For example, the leaching of sphalerite in the presence of ferric ions occurs by oxidative dissolution:

$$ZnS + 2Fe^{3+} \longrightarrow Zn^{2+} + 2Fe^{2+} + S \tag{3.1}$$

While the leaching of sphalerite in an acidic solution in which no oxidant is present occurs by nonoxidative dissolution:

$$ZnS + 2H^+ \longrightarrow Zn^{2+} + 2H_2S \tag{3.2}$$

Both the oxidative and nonoxidative dissolution of minerals are electrochemical processes, since charged species move across the phase boundary.

In this chapter, the electrochemical model of leaching is discussed, the kinetic equations derived and their application in both leaching and bacterial leaching studies demonstrated.

3.2 Electrochemical Mechanism of Oxidative and Reductive Reactions

3.2.1 Anodic and cathodic reactions

Dissolution reactions that result in the change of oxidation state of the mineral can be separated into anodic and cathodic half-reactions. For example, during the dissolution of a metal, M, in an aqueous solution containing the oxidant, B^{2+}, the dissolution reaction is:

$$M(s) + B^{2+}(aq) \longrightarrow M^+(s) + B^+(aq) \tag{3.3}$$

This reaction can be separated into the anodic dissolution of metal, M, and the cathodic reduction of oxidant, B^{2+}. The anodic half-reaction is:

$$M(s) \longrightarrow M^+(aq) + e(s) \tag{3.4}$$

and the cathodic half-reaction is given as Eq. 3.5:

$$B^{2+}(aq) + e(s) \longrightarrow B^+(aq) \tag{3.5}$$

Note that the separation of the reaction into anodic and cathodic half-reactions is a formal separation and that it does not mean that the reaction necessarily occurs at different locations on the mineral surface.❶ The anodic and cathodic half-reactions occur at all sites on the particle surface, unless the mineral particle is clearly heterogeneous as in the case of galvanic corrosion.

Since the rate of production of electrons is equal to the rate of consumption of electrons, the general requirement for dissolution is that the sum of the rates of the anodic reactions must be equal to the sum of the rates of the cathodic reactions. This condition is given as:

$$r_a = r_c \tag{3.6}$$

where r_a and r_c are the rates of the anodic and cathodic half-reactions, respectively. This condition is used to derive an expression for the potential difference at the mineral surface and the rate of dissolution.

The rate of an electrochemical reaction is dependent on the potential difference across the electrode-solution interface, E, the concentration of reacting species and the temperature, T. The potential, E, is that presented earlier as ψ_{13}, but is measured with respect to a reference electrode, that is, $E = \psi_{13} - \psi$.

If the anodic half-reaction is considered irreversible, then the rate of the anodic half-reaction is given as:

$$r_a = k_a \exp[\alpha_a FE/(RT)] \tag{3.7}$$

where k_a is a rate constant, α_a is the charge-transfer coefficient, which has a value between 0.4-0.6, F is Faraday's constant and R the gas constant. Similarly, the rate of the cathodic half-reaction is given as:

$$r_c = k_c[B^{2+}] \exp[-(1-\alpha_c)FE/(RT)] - k_c^1[B^+]\exp[\alpha_c FE/(RT)] \tag{3.8}$$

where k_c and k_c^1 are rate constants and α_c is the charge-transfer coefficient for the cathodic reaction which also has a value between 0.4-0.6.

Substituting the expressions for r_a and r_c (Eq. 3.7 and Eq. 3.8), into Eq. 3.6 gives the expression:

$$k_a \exp[\alpha_a FE/(RT)] = k_c[B^{2+}] \exp[-(1-\alpha_c)FE/(RT)] - k_c^1[B^+]\exp[\alpha_c FE/(RT)] \qquad (3.9)$$

If it is assumed that $\alpha_a = \alpha_c$, then Eq. 3.9 may be rearranged to give an expression for the potential difference, E:

$$E = \frac{RT}{F} \ln\left(\frac{k_c[B^{2+}]}{k_a + k_c^1[B^+]}\right) \qquad (3.10)$$

This potential is called the mixed potential and as a result the electrochemical mechanism of leaching is sometimes referred to as the mixed-potential model. Substituting this expression for E back into Eq. 3.6 gives the following equation for the rate of dissolution:

$$r_{diss} = k_a \left(\frac{k_c[B^{2+}]}{k_a + k_c^1[B^+]}\right)^\alpha \qquad (3.11)$$

where $\alpha_a = \alpha_c = \alpha$ and r_{diss} represents the rate of dissolution, which is equal to the rate of the anodic or the cathodic half-reactions, i.e., $r_{diss} = r_a = r_c$ by Eq. 3.6. The value of α is expected to be close to 0.5, so this model indicates that the rate of dissolution is expected to be one-half order in the concentration of the oxidant, B^{2+}. Many dissolution reactions display a one-half order dependence on the concentration of the oxidant in solution.

The derivation of Eq. 3.11 represented a major advance in the understanding of leaching chemistry. It was first presented by Nicol et al. for the leaching of UO_2 in ferric sulfate solutions. The kinetics of reductive leaching, such as the reduction of pyrolusite (MnO_2) by sulfur dioxide can be described by a similar mechanism. It is easy to show that the rate of dissolution for a reductive leaching reaction is given by:

$$r_{diss} = k_c \left(\frac{k_c[B^+]}{k_c + k_a^1[B^{2+}]}\right)^{1-\alpha} \qquad (3.12)$$

The value of α is expected to be close to 0.5, so that order of reaction with respect to the reductant is expected to be close to 0.5. A large number of leaching studies have been published and of particular significance is the result that the order of reaction is close to 0.5 for many oxidative and reductive dissolution reactions. This indicates that the electrochemical mechanism of leaching is generally applicable.

3.2.2 Influence of the electronic structure of the mineral on the dissolution rate

The rates of leaching of a mineral from different sources are different. For example, the rate of dissolution of sphalerite from different ores varies depending mainly on the concentration of iron in the sphalerite lattice. This phenomenon is explained by the influence that the electronic structure of the mineral exerts on the electrochemistry of dissolution. In order to dissolve a mineral, electrons need to be removed from bonding orbitals. In the case of sphalerite it is energetically more favorable to remove electrons from the bonding orbitals of iron in the lattice than from zinc in the lattice.

Different minerals dissolve at different rates. For example, it is well known that pyrite dissolves slowly in comparison to other iron sulfides, such as pyrrhotite. These differences can also be explained by the semiconductor-electrochemical model of leaching. The electronic structure of pyrite is such that the upper valence band, which interacts with common oxidizing agents, does not contribute to the bonding of the mineral. Therefore, removal of electrons from these nonbonding orbitals will not result

3.3 Applications of Electrochemical Mechanism in Leaching

3.3.1 Chemical leaching of pyrite by ferric ions

Pyrite is a common mineral that is found in many or ore deposits and tailling sites. The primary cause of acid-mine drainage is the bacterial leaching of the pyrite in tailing sites. Gold that is encapsulated by pyrite and arsenopyrite cannot be treated by the conventional cyanidation process. Pretreatment of these ores by bacterial leaching liberates this gold. Thus, pyrite plays an important role in the study of the mechanisms of bacterial leaching and the study of the electrochemical mechanism of dissolution of pyrite will assist in the process of understanding bacterial leaching.❷

McKibben and Barnes found that the rate of leaching of pyrite in ferric sulfate solutions is given as:

$$r_{diss} = k \frac{[Fe^{3+}]^{0.5}}{[H^+]^{0.5}} \tag{3.13}$$

These results may be explained in terms of the electrochemical mechanism of dissolution. If it is assumed that the initial step is the formation of a hydroxide layer on the surface of the pyrite and the rate-determining step is the electrochemical reaction of this layer, then we can write the first two steps in the anodic dissolution of pyrite as follows:

$$FeS_2 + H_2O \rightleftharpoons FeS_2OH^-_{ads} + H^+ \tag{3.14}$$

$$FeS_2OH^-_{ads} \longrightarrow FeS_2OH_{ads} + e \tag{3.15}$$

The chemical and electrochemical reactions that follow these two initial reaction are assumed to be much faster than the rate-determining reaction and therefore do not influence the rate of reaction. The rate of formation of the adsorbed species $FeS_2OH^-_{ads}$ by Eq. 3.14 is given by:

$$r_1 = k_1(1-\theta) - k_{-1}\theta[H^+] \tag{3.16}$$

where q is the fraction of surface covered by $FeS_2OH^-_{ads}$. Since this reaction is fast compared to the subsequent rate-determining step and there is no accumulation of the adsorbed layer of $FeS_2OH^-_{ads}$, then $r_1 \approx 0$. Assuming that the surface coverage θ is small ($\theta \ll 1$), we obtain the following expression for θ:

$$\theta = \frac{k_1}{k_{-1}[H^+]} \tag{3.17}$$

The rate of the rate-determining electrochemical step (Eq. 3.15) is given by:

$$r_a = k_a\theta \exp[\alpha_a FE/(RT)] \tag{3.18}$$

The substitution of Eq. 3.17 into Eq. 3.18 gives:

$$r_a = \frac{k_a k_1 \exp[\alpha_a FE/(RT)]}{k_{-1}[H^+]} \tag{3.19}$$

The cathodic reaction is the reduction of ferric ions at the pyrite surface, given by:

$$Fe^{3+} + e \longrightarrow Fe^{2+} \tag{3.20}$$

The rate of the cathodic reaction is given by:

$$r_c = k_c[Fe^{3+}]\exp[-(1-\alpha_c)FE/(RT)] \tag{3.21}$$

Following the same method of derivation described in the section on the electrochemical model of leaching (i.e., using Eq. 3.6 to get an expression for E, and substituting this expression into the rate equation for the anodic reaction), the following expression for the dissolution of pyrite is obtained:

$$r_{diss} = \left(\frac{k_a k_1}{k_{-1}[H^+]}\right)^{1-\alpha}(k_a[Fe^{3+}])^\alpha \tag{3.22}$$

where $\alpha = \alpha_a = \alpha_c$.

Since α_a and α_c are expected to have values close to 0.5, it is clear that this rate equation is the same as that obtained by McKibben and Barnes. This means that the electrochemical model describes the kinetics of dissolution of pyrite.

3.3.2 Effect of bacterial action on the mixed potential of pyrite

In the derivation of the electrochemical mechanism of leaching, we obtained an expression for the mixed potential, given by Eq. 3.10. The mixed potential of a dissolving mineral can be measured with respect to a reference electrode. This is an easy and useful method of determining the effect that changes in the solution will have on the leaching reaction.[3] Crundwell et al. measured the mixed potential of a pyrite electrode in the presence and absence of bacteria to determine the influence of biofilm formation on the kinetics of bacterial leaching. The results of this mixed potential study are shown in Fig. 3.1.

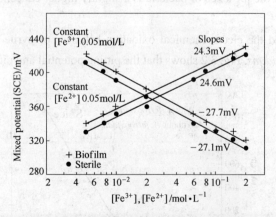

Fig. 3.1 The mixed potential of pyrite in the presence and absence of bacteria at various concentrations of ferrous and ferric ions

The experiments shown in Fig. 3.1 ware conducted in two steps. Initially, the pyrite electrode was pretreated in a leaching solution in either the presence or absence of bacteria for 10 days. After this process the electrode was removed from the solution and the mixed potential was determined at different concentrations of ferrous and ferric sulfate in a sterile solution containing 0.4mol/L sodium sulfate rid 0.2mol/L sulfuric acid. A slight blackening of the surface was observed in experiments in the presence of bacteria. This was thought to be caused by the formation of a ferric hydroxide layer on the pyrite surface. In addition, micro-colonies of bacteria or biofilms, were clearly visible by the end of the pretreatment process.

The results shown in Fig. 3.1 indicated that there was an increase in the mixed potential of about 10mV when bacteria were present in the solution. This slight increase in the mixed potential may be explained by a slight increase in the concentration of ferric ions at the surface as a result of the presence of the bacterial biofilm. However, from an examination of Eq. 3.7, the increase in the leaching rate as a result of an increase in the mixed potential of 10mV can be no greater than 22% at 25°C (i.e., $\exp[\alpha 0.01 F/(RT)] = 1.22$. Such a small increase in the rate of leaching as a result of bacterial activity suggests that the contribution of bacterial attachment to the overall rate of leaching is small.

These results are explained by the electrochemical mechanism of leaching. If it is assumed that the partial anodic current due to the dissolution of pyrite is small compared to the partial currents due to the oxidation and reduction of the ferric and ferrous ions in solution, that is $k_a \ll k_c^1[Fe^{2+}]$, then the mixed potential Eq. 3.10 may be written as:

$$E = E_{pyr} + \frac{RT}{F} \ln\left(\frac{[Fe^{3+}]}{[Fe^{2+}]}\right) \qquad (3.23)$$

where E_{pyr} is $RT/F \ln(k_c/k_c^1)$.

The slopes presented in Fig. 3.1 indicate that the slope of the mixed potential with respect to $\ln[Fe^{3+}]$ and $\ln[Fe^{2+}]$ are close to 26.5mV, i.e., the value of RT/F at 25°C. These results indicate that the effect of changing the concentrations of ferrous and ferric ions on mixed potential in the presence and absence of bacteria is described by Eq. 3.23. This suggests that the most likely reason for the slight increase in the mixed potential in the presence of bacteria is a slightly higher concentration of ferric ions at the pyrite surface.

Gu et al. investigated the electrochemical oxidation behavior of pyrite in bioleaching system of *Acidthiobacillus ferrooxidans*. Fig. 3.2 shows that the pitting potential and pitting current density of the

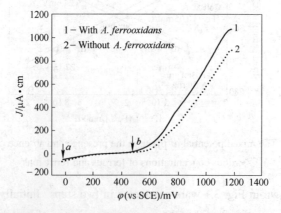

Fig. 3.2 Polarization curves of pyrite in the presence and absence of *A. ferrooxidans*
(pH = 2.0; t = 30°C; scan rate 1mV/s)

pyrite electrode have no significant difference with or without bacteria involvement. But the polarization current density of pyrite in the leaching system with *A. ferrooxidans* was higher than that in the sterile solution after the potential reaches the pitting potential. That is to say, the corrosion rate of pyrite was accelerated by *A. ferrooxidans*. In the presence or absence of *A. ferrooxidans*, the oxidation reaction of pyrite is also divided into two steps: the first reaction step involves the oxidation of pyrite to S, and the second reaction step is the oxidation of S to SO_4^{2-}. The oxidation mechanism of pyrite is not changed in the presence of *A. ferrooxidans*, but the oxidation rate of pyrite is accelerated.

3.3.3 Effect of electrochemical bioleaching on the copper recovery

Ahmadi et al. investigated the conventional and electrochemical bioleaching were to extract copper from Sarcheshmeh chalcopyrite concentrate at high pulp densities. Experiments were conducted in the presence and absence of a mixed culture of moderately thermophilic iron- and sulphur oxidizing bacteria using a 2-L stirred electrobioreactor at 20% (w/v) pulp density, an initial pH of 1.4–1.6, a temperature of 50°C, a stirring rate of 600r/min and Norris nutrient medium with 0.02% (w/w) yeast extract addition. Fig. 3.3 shows the results of 10 day leaches showed that, when using electrochemical bioleaching in an ORP range of 400mV to 430mV, copper recovery reaches about 80% which is 1.5 times higher than that achieved in conventional bioleaching. It appears that applying current directly to the slurry optimizes both, the biological and chemical subsystems, leading to an increase in both, the dissolution rate and the final recovery of copper from the concentrate.

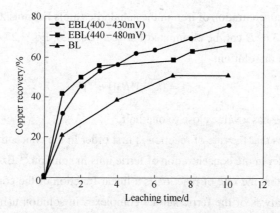

Fig. 3.3 Copper recovery as a function of leaching time for bioleaching (BL) and electrochemical bioleaching (EBL)

3.4 Electrochemical Kinetics and Modeling

3.4.1 Kinetics of the oxidative dissolution of sphalerite

The dissolution of sphalerite in ferric sulfate or ferric chloride solutions occurs according to the reaction:

$$ZnS + 2Fe^{3+} \longrightarrow Zn^{2+} + 2Fe^{2+} + S \qquad (3.24)$$

Iron occurs as an impurity in all natural sphalerite. This iron has a pronounced effect on the rate of dissolution in ferric solutions. Crundwell determined that the rate of dissolution is proportional to the iron content in the sphalerite. Values of L/τ_{rxn}, which is directly proportional to the rate of dissolution. (L is the initial particle size, and τ_{rxn} is the time for complete conversion of particles initially of size L).

The anodic half-reaction for the dissolution of sphalerite is written as:

$$ZnS \longrightarrow Zn^{2+} + S + 2e \tag{3.25}$$

and the cathodic half reaction is given as:

$$2Fe^{3+} + 2e \longrightarrow 2Fe^{2+} \tag{3.26}$$

The presence of iron that substitutes for zinc atoms in the sphalerite lattice results in a d-orbital band within the band-gap of sphalerite. The iron d-orbitals of this band are of bonding character. This means that the removal of an electron from this band also results in dissolution of the solid. The iron impurity and its associated d-orbital affects the dissolution of sphalerite by presenting a narrow localized band with which the transfer of electrons is energetically more favorable than with the valence band.

This means that the rate of the anodic reaction is given as:

$$r_a = k_a N_d \exp[EF/(RT)] \tag{3.27}$$

and the cathodic reaction is given as:

$$r_c = k_c N_d [Fe^{3+}] \exp[-(1-\alpha_c)FE/(RT)] \tag{3.28}$$

where N_d is the relative concentration of iron in the sphalerite (mols Fe/mols Zn).

Since $r_a = r_c$ by Eq. 3.6, E can be eliminated between Eq. 3.6, Eq. 3.27 and Eq. 3.28, giving the expression for the rate of dissolution:

$$r_{diss} = (k_a k_c)^\alpha N_d [Fe^{3+}]^\alpha \tag{3.29}$$

where $\alpha = \alpha_a = \alpha_c$, and α has a value close to one-half.

This equation indicates that the rate of reaction is first order in the concentration of iron in the sphalerite and is one-half order in the concentration of ferric ions in solution.[4] Experimental results shown in Fig. 3.4 demonstrate that the rate of reaction is a linear function of the concentration of iron in the lattice. In a detailed analysis of the ferric sulfate complexes in solution using the stability constants calculated by Dry, Crundwell argued that of the ferric sulfate species in solution only $FeHSO_4^{2+}$ and $Fe^{3+}(aq)$ were active at the sphalerite surface.

The rate of reaction is then:

$$r_{diss} = (k_a k_c)^{0.5} N_d ([Fe^{3+}(aq)] + [FeHSO_4^{2+}])^{0.5} \tag{3.30}$$

The rate of reaction, r_{diss}, is proportional to $1/\tau_{rxn}$ where τ_{rxn} is the time complete reaction of the particle. Values of $1/\tau_{rxn}$ are plotted as a function of the concentration of the active ferric species in solution in Fig. 3.4. Since the slope of the regression line of data is 0.45, which is close to 0.5, this figure indicates that Eq. 3.30 is in agreement with the experimental results for the kinetics of the oxidative dissolution of sphalerite.

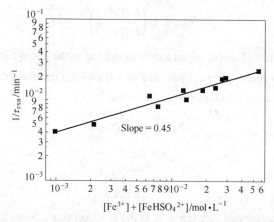

Fig. 3.4 The effect of solution speciation on the rate of leaching
($1/\tau$ is proportional to the rate of leaching)

The discussion indicates that the electrochemical mechanism is able to describe the two most important features of the leaching of sphalerite: that the rate of reaction is a linear function of the concentration of iron in the lattice and the rate is one-half order in the concentration of oxidant in solution.

3.4.2 Electrochemical kinetics and model of bacterial leaching

The kinetics of bacterial leaching reactions have not been generally interpreted in terms of the electrochemical mechanism of leaching. Part of the reason for this is that in typical bacterial leaching studies the concentrations of ferric and ferrous ions vary during the course of the experiment. This makes it difficult to analyze the data to obtain the orders of reaction that are required to identify the reaction as being controlled by the electrochemical step. In order to undertake an analysis of batch data of this sort, it is necessary to present the electrochemical mechanism in the form of a mass balance describing the leaching reaction in a batch reactor.

As an example, the leaching of sphalerite by ferric sulfate is examined. This derivation can be easily applied to the leaching of other minerals. The mass balance for the leaching of sphalerite in a batch reactor is given as:

$$\frac{dN_{ZnS}}{dt} = -r_{diss}A \tag{3.31}$$

where N_{ZnS} represents the number of moles of sphalerite in the batch reactor, r_{diss} represents the rate of consumption of sphalerite by the leaching reaction (units of mol/(m²·s)), and A represents the surface area (m²) of sphalerite that is available for reaction.

To develop the mathematical model expressions for the rate of reaction, r_{diss}, and for the available surface area, A, are required. These two expressions are derived from the electrochemical mechanism of leaching and from the shrinking particle model.

An expression for the rate of reaction has been presented in Eq. 3.30. Although Crundwell showed that not all of the ferric species are active at the sphalerite surface, it is assumed that all the ferric species are active at the surface in order simplify this presentation. Thus the rate of dissolution is given by Eq. 3.11, and if it can be assumed that $k_a \ll k_c^1[Fe^{2+}]$, then:

$$r_{\text{diss}} = k_a \left(\frac{k_c [\text{Fe}^{3+}]}{k_c^1 [\text{Fe}^{2+}]} \right)^\alpha \tag{3.32}$$

The change in the number of moles of sphalerite is related to the change in size of the particle. For particles that are approximately spherical with radius 1 (or which maintain their geometric proportion during leaching), the following relationships are obtained:

$$dN_{\text{ZnS}} = 4\pi \lambda^2 \rho_{\text{ZnS}} d\lambda \tag{3.33}$$

$$A = 4\pi \lambda^2 \tag{3.34}$$

$$X = 1 - \left(\frac{\lambda}{L} \right)^3 \tag{3.35}$$

where N_{ZnS} is the molar density of sphalerite, X is the reaction conversion, and L is the initial particle size.

Substituting Eq. 3.32-Eq. 3.35 into the mass balance, Eq. 3.31, and after some rearrangement, the mass balance for the leaching of sphalerite in a batch reactor can be written as:

$$\frac{dX}{dt} = \frac{3}{\tau_{\text{rxn}}} (1-X)^{2/3} \left(\frac{[\text{Fe}^{3+}]}{[\text{Fe}^{2+}]} \right)^\alpha \tag{3.36}$$

where τ_{rxn}, the time for complete reaction, is given by $\tau_{\text{rxn}} = \frac{\rho_{\text{ZnS}} L}{k_a} \left(\frac{k_c^1}{k_c} \right)^\alpha$.

The physical significance of τ_{rxn} is that it represents the time that a particle of size L will take to be completely dissolved if the concentration of ferric and ferrous ions remains at the initial value throughout the course of the reaction.

The effect of temperature is accounted for by the Arrhenius equation, $\tau_{\text{rxn}} = \tau_{\text{rxn}}^\infty \exp[E_A/(RT)]$. The initial condition is $X = 0$ at $t = 0$.

Eq. 3.36 is solved in conjunction with the solution mass balances, given as:

$$[\text{Fe}^{3+}] = [\text{Fe}^{3+}]_0 - 2F_{\text{ZnS}} X - 2F_{\text{FeS}} X \tag{3.37}$$

$$[\text{Fe}^{2+}] = [\text{Fe}^{2+}]_0 + [\text{Fe}^{3+}]_0 - [\text{Fe}^{3+}] + F_{\text{FeS}} X \tag{3.38}$$

where F_{ZnS} and F_{FeS} represent the initial molar concentrations of zinc sulfide and iron sulfide added to the batch reactor.

The fit of the model parameters to the leaching data presented by Verbaan and Crundwell is shown in Fig. 3.5. The sphalerite contained 50.9% zinc, and 9.1% iron and the initial particle size was 22pm. The model is represented the lines and the data by the points on these figures.

The close correspondence between the model (Eq. 3.36-Eq. 3.38) and the data suggests that the model is a good representation of the data and supports the assumption that, for this data, all ferric species are active in the reaction. All the lines in both these figures have been obtained with the same set of parameters, that is, with $3/\tau_{\text{rxn}}^\infty = 1.88 \times 10^5$ min^{-1}, $\alpha = 0.39$, and $E_a = 48$kJ/mol. These parameters are similar to those obtained from other studies of chemical leaching. The value of 0.39 obtained for α is close to the value predicted from the electrochemical mechanism. Therefore, it can be concluded that the leaching of sphalerite is described by the electrochemical mechanism.

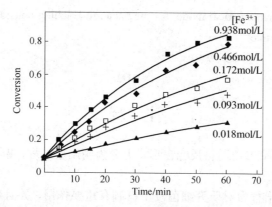

Fig. 3.5 Effect of different concentrations of ferric ions on the rate of leaching of sphalerite at 65°C, 0.038mol/L Fe^{2+}, 0.1mol/L H_2SO_4, 1g/L solids

The parameters obtained from batch leaching studies can be used to predict the operation of a continuous leaching plant. The piloting of a new continuous operation is a slow and expensive task. In contrast, small-scale batch leaching experiments are easy to perform.❺ Therefore, a thorough analysis of the batch leaching data in the manner shown in this chapter and the use of the kinetic parameters obtained from this analysis to predict the performance of a continuous plant could save a significant amount of time and money in feasibility and piloting studies. In addition, these kinetic parameters can be used in models to simulate and optimize the operation of the full-scale continuous plant.

3.5 Summary

Dissolution reactions that result in the change of oxidation state of the mineral can be separated into anodic and cathodic half-reactions. The electrochemistry of bacterial leaching revolves around the electrochemical model of dissolution, and its derivation has been presented in detail. The electrochemical model has been shown to hold for three different dissolution reactions and demonstrates much promise for the interpretation of the kinetics of bacterial leaching. The electrochemical mechanism of leaching is assumed that the partial anodic current due to the dissolution of pyrite is small compared to the partial currents due to the oxidation and reduction of the ferric and ferrous ions in solution: these indicate that the effect of changing the concentrations of ferrous and ferric ions on mixed potential in the presence and absence of bacteria. This suggests that the most likely reason for the slight increase in the mixed potential in the presence of bacteria is a slightly higher concentration of ferric ions at the pyrite surface.

Questions

1. What is a dissolution process that will affect the rate of solid phase dissolution into liquid phase?
2. How does the electronic structure of the mineral influence the rate of dissolution?
3. How does pyrite chemical leaching work by ferric ions?
4. What is the dissolution equation for a reductive leaching reaction?

5. How to develop a mathematical model for the bacterial leaching considering the electrochemical kinetics, and give an example.

Note

注　释

❶ 请注意，这里为研究方便而将反应在形式上分为阴阳极过程，并不是表示两个在矿物不同位置发生的反应。

❷ 研究黄铁矿的浸出行为对研究细菌浸出机制有重要作用，并且研究黄铁矿溶解的电化学机制也有助于对细菌浸出的理解。

❸ 式 3.10 是通过对浸出电化学机制的推导得出的混合电势表达式。矿物的溶解电势可以通过参比电极测量，并用此法研究溶液变化对浸出反应式的影响。

❹ 本式表明反应率是由闪锌矿中铁含量和溶液中的铁离子浓度决定的。

❺ 一个新的连续操作参数的制定需要花费较长时间和较大成本，而分批浸出获得的参数可以用于指导连续浸出的操作参数设置。

4 Biohydrometallurgy of Copper

本章要点 采用矿坑或钻孔可直接在原位置进行浸矿提取金属铜。在铜的溶解过程中，铁离子是主要的氧化剂，而利用生物浸矿过程产生的铁离子是其优势之一。铜的化学浸矿初期受酸的添加量影响，但后期为保证细菌的正常生长必须保持稳定合理的 pH 值；从酸溶液中回收铜经典的方法是采用铁离子凝合法。电解提取法利用系列电解槽循环使用电解质，并采用铅合金阳极、不锈钢阴极或铜始极片。

4.1 Introduction

The application of the bioleaching reaction for copper has been exploited and used to develop suitable methods to recover copper from copper-bearing solutions.

The heap leaching of copper has been practiced for several decades, mostly with oxide ores. Bioleaching of copper is of major importance in metals production and this chapter will look at recent operations and the innovations that are making economic recovery of secondary sulfide copper possible. Before examining the application of bioleaching to copper recovery, it is necessary to have a review on the basic mineralogy and some recent ideas that are providing a new viewpoint on the chemistry of leaching.

We lack a complete understanding of the various mechanisms and are dealing with living organisms that have somewhat unpredictable behavior.

This chapter is intended as a short review of copper bioleaching, its application, bioleaching experience and some examples of existing operating plants. No doubt the advance of learning and its application will create more economically viable and efficient processing technology.

4.2 Definitions and Mineralogy Related to Copper Leaching

One of the most confusing aspects in reviewing copper leaching is a confusion of terms used for each type of leach operation. Therefore, for the purposes of this chapter the following terminology will be used:

(1) Dump leaching is leaching from existing run-of-mine leach dumps which were previously considered waste and includes run-of-mine ore placed specifically for leaching. This type of leaching does not use size reduction equipment and relies on mine blasting for size reduction.

(2) In situ leaching is the leaching of ore in place without removal from the ore body, using adits or drill hole solution systems.❶

(3) Heap leaching refers to crushed ores placed on prepared pads in layers for leaching.

(4) Permanent pad refers to a prepared leach pad that is used continuously without removal of the leached ore prior to stacking of fresh ore.

(5) Dynamic or ON/OFF pad refers to prepared pads from which the ore is removed prior to placing fresh ore for leaching.

The main confusion occurs with reference to the terms "dump" and cheap leaching. Dump leaching is carried out on untreated uncrushed run-of-mine material whereas heap leaching involves crushed, pretreated ore.

This short glance at the mineralogy of copper minerals is not intended to be complete as there are over 350 copper minerals. What will be discussed are the most common copper sulfide species considered suitable for bioleaching. Also included is the oxidation of pyrite which forms a fundamental part of copper bioleaching.

4.2.1 Pyrite

A short review of pyrite oxidation is appropriate since these reactions are the driving force behind what happens mineralogically inside heap and dump leaching operations. The oxidation of iron disulfide to the sulfate ion involves the transfer of as many as 16 electrons with the potential formation of complex intermediate species. The overall reaction takes place at pH<4 in heaps and dumps and releases about 1,500kJ of heat in the following reaction:

$$FeS_2(s) + 14Fe^{3+}(aq) + 8H_2O \longrightarrow 15Fe^{2+}(aq) + 2SO_4^{2-}(aq) + 16H^+ \qquad (4.1)$$

The reaction rate is fast, but limited in the absence of assistance from bacteria such as *Thiobacillus ferrooxidans* which catalyzes the reoxidation of ferrous to ferric iron as shown below:

$$bacteria + 4Fe^{2+}(aq) + O_2(g) + 4H^+ \longrightarrow 4Fe^{3+}(aq) + 2H_2O \qquad (4.2)$$

4.2.2 Secondary sulfides

The most common secondary copper mineral considered in heap bioleaching are chalcocite and covellite. There is generally a local participation of ferric iron in the process derived from dissolution of either pyrite or chalcopyrite. The earliest work by Sullivan in 1930 indicated that chalcocite dissolved in two stages according to the reactions:

$$Cu_2S + 2Fe^{3+} \longrightarrow Cu^{2+} + 2Fe^{2+} + CuS \text{ (not the mineral covellite)} \qquad (4.3)$$

$$CuS + 2Fe^{3+} \longrightarrow Cu^{2+} + 2Fe^{2+} + S^0 \qquad (4.4)$$

In 1967, Thomas et al. leached both synthetic chalcocite and digenite and found that chalcocite was converted to digenite and thereafter both samples were converted into $Cu_{1.1}S$, at that time considered to be one "mineral" -blaubleibender covellite. The proposed system from this discovery was identical to the reaction given by Sullivan but with quite a different reaction path.

More recently, Gobleld published a study of the leaching of chalcocite with Fe^{3+}(aq) and found that an entire series of nonstoichiometric copper sulfides were produced. From this work a variety of possible chalcocite-series leaching sequences are possible (recent work in leaching of chalcocite in bacterial and sterile conditions has shown that the bacteria do increase the rate of reaction and can be involved in these reactions). Possible chalcocite leaching sequences are as follows:

$$Cu_2 \to Cu_{1.97}S \to Cu_{1.8}S \to Cu_{1.75}S \to Cu_{1.6}S \to Cu_{1.4}S \to Cu_{1.12}S \to CuS$$

chalcocite dijurleite digenite anilite geerite spionkopite yarrowite covellite

$$Cu_{1.78}S \to Cu_{1.74}S \to Cu_{1.12}S \quad \Rightarrow \quad \text{common reactions}$$

roxbyite "digenite" yarrowite $\to$ rare reaction

4.2.3 Primary sulfides

The major copper primary sulfide considered economically important is chalcopyrite. Chalcopyrite is bioleachable, but is currently considered to be nonviable economically due to the long leach times required. Some operations have used waste dumps, irrigated over a four to six year period to recover copper from chalcopyrite. Reported recoveries are only about 15%. Considerable effort is being expended to improve recovery of primary sulfides through bioleaching in heaps. Most work has been on the bioleaching of concentrates rather than mined ore. Much of the concentrate leaching uses stirred tank reactors; this has not been reviewed in this chapter since these processes have not yet been applied commercially.

4.3 Physico-Chemical Leaching Variables

Most of the variables that effect leaching rates and recoveries are common to heap, dump, or in situ leaching systems. These variables can be divided into those affecting the chemical part of leaching and those that affect the bacterial component. The primary variable is the mineralogy of the ore and it is an assumed variable. The remaining variables are discussed in the following three sections.

4.3.1 Surface area

The rock size used for leaching has a direct effect on the leach rate. Whether produced by pit blasting or by crushing, rock size is specific to each ore deposit and depends on the mineral occurrence and whether it is disseminated or in boundary layers within the rock. The effect of particle size must be tested for each ore type. In general, the finer the particle size the better the final recovery. It is interesting to note that frequently, initial recovery is faster with ore of a larger size due to the reduced surface area, although the final recovery is adversely affected. In situ leaching is primarily affected by the mineral formation and fissure sizes.

The chemical leach component is affected by the ability of the solution to contact the copper and to transport the dissolved copper out of the rock fissures. Work on the use of a wetting agent to reduce solution surface tension to allow better fissure penetration has taken place, especially with oxide ores. Little work of this type has been carried out on bacterial leaching of sulfide minerals. In general it is felt that wetting agents or surfactants are detrimental to the bacteria because of the potential of these agents to rupture bacterial cells.

4.3.2 Acid levels

In dump leaching, a preconditioning step with high acid levels is frequently employed. The benefit of this increased acid level is specific to each ore type. Too much acid will increase the overall acid usage due to acid consumption by the gangue. In heap leach operations, agglomeration is commonly practiced. This agglomeration step releases some chemically available copper. The acid addition rate during the agglomeration step is ore-specific. As with dump leaching pretreatment, the initial acid level for agglomeration must be balanced against gangue mineral consumption and the acid required to optimize bacterial activity.

4.3.3 Oxidants

Ferric ions are considered to be the primary oxidant in the dissolution of copper and it is the production of ferric ions that is the main benefit of bioleaching. A possible alternative to bioleaching would be the physical addition of ferric iron.[2] However, this is an expensive alternative unless there is a means of reoxidizing the ferrous iron. There are several patents on the regeneration of ferric iron, but none of these alternatives are as cost-effective as bacterial oxidation. Copper grades sufficiently high to carry the cost of copper recovery by chemical leaching are not considered in this chapter. Other potential oxidants cannot be used in conjunction with bacterial leaching since they have a detrimental effect on the bacteria.

4.3.4 Agglomeration

A common innovation in heap leach operations is to mix the ore with acid and water to form an agglomerate. This agglomerate differs from traditional gold or concentrate agglomerates in that a true 'pellet' is not formed. The purpose of agglomeration is to prevent the segregation of fine and coarse material during stacking. The moisture level within the agglomerate also partially determines the permeability of the stacked material.

Agglomeration takes place either in rotating drums (Fig. 4.1), reversing conveyor belts or conveyor

Fig. 4.1 Agglomeration drums at the Quebrada Blanca Operation, Chile

belts with a series of plows to mix the ore and water/acid. Most newer operations have included rotating drums during initial design, but in some cases these have been fitted later to improve agglomerate quality. Typical agglomerate discharge moisture is about 10% by weight. A larger crushed ore size will decrease the required moisture levels to form a suitable agglomerate. Hot water can be used for agglomeration to add heat to the leach pile and start more rapid leaching.

An interesting phenomenon associated with agglomeration is that some copper is liberated quickly as a result of this process. Leaching an ore in the laboratory with acid does not liberate the same quantity of copper as does the agglomeration process. The use of a binder to improve permeability of the agglomerate may be considered in the future, but there are no reports of the application of binders in copper heap leaching.

4.3.5 Curing time

The time required for the acid and moisture to act on the minerals must be considered. In heap leach operations using agglomeration, the time after agglomerating, prior to irrigation is considered important. At the LoAguirre copper operation three days is specified, at Cerro Colorado several weeks is allowed and at the Quebrada Blanca Operation the curing time is not considered an important variable. Tests must be carried out on each ore type to determine the appropriate curing time.

4.3.6 Permeability

The permeability of a heap or dump helps to determine the solution distribution and ingress of oxygen required for bacterial activity. Solution ponding on top of the heap must be avoided since this will inhibit oxygen penetration. The permeability will decrease as a dump or heap ages and solution irrigation rates may have to be reduced. A system of ON/OFF irrigation may also be employed in older leach areas. Good agglomeration greatly improves permeability and consistency on the pile and prevents solution channeling.

4.4 Bacterial Leaching Variables

Many of the variables discussed above also have a bearing on bacterial activity. The available surface area effects bacterial growth and solution transport, as does the level of acid and other oxidants. The heap permeability has a direct effect on oxygen availability for bacterial growth. The more important variables affecting bacterial activity are discussed below.

4.4.1 Acidity

The amount of acid added initially effects the chemical leaching of copper, but suitable pH levels for bacterial growth must be maintained.[3] Desirable pH levels are 1.8 to 2.2. Typical solution acid concentrations are 6.0-8.0g/L sulfuric acid. In heap or dump leaching, consideration must be given to the change in acid concentrations from the top to the bottom of a leach pile (i.e., the pH rises as acid is consumed while the solution drains through the pile). This fact makes acid control difficult at times and influences the height of heap construction. Adaptation of the bacteria helps this situation, and if there is sufficient oxygen, there will be a bacterial front moving through the pile.

4.4.2 Oxygen

A theory of bacterial growth in a heap holds that the major area of bacterial growth is in the top 1.5m of a leach pile due to the limits of oxygen diffusion into the surface of a heap. Measurements on nonaerated heaps have shown that oxygen levels drop after the first 1.5m from the top of the leach pile to below 5% oxygen at the bottom. To improve natural diffusion of oxygen, cyclical irrigation has been used. The theory of cyclical irrigation is that the draining of the irrigation solution will result in oxygen being drawn into a leach pile. No data is available to demonstrate this theory.

In dump leach operations the natural segregation of coarse and fine ore as the dump is constructed allows air to ingress into the bottom of a leach dump. Air ingress has been further improved by the "finger dump" design. The idea of this design is to optimize the "chimney effect" within the dump. The effect of drilled air holes within dumps and the addition of air by use of fans have been also been tested. Finger dump construction is the norm for leaching operations and forced air addition has been limited to test work.

The situation for heap leaching is somewhat different. The finer crushed ore does not allow for the same amount of coarse and fine material segregation and this segregation is not desirable for permeability reasons. In heap leaching, air diffusion theory can be accepted for heap operation, but improved leach and recovery rates are possible. Recent developments in bacterial heap leach operations have seen the addition of air injection systems to reduce leach cycle times. Initial air addition tests used the bottom drain systems to inject air. In at least one operation a separate air injection system that uniformly distributes air through the heap has been installed. The results show that aeration which is achieved with air injection on a large scale operation can reduce leach cycle times. The effects on overall copper recovery are not clearly defined since this plant achieved high copper recovery without aeration, although with longer each cycle times.

The addition of air to a heap leach operation is accompanied by several changes. When air is first added, iron levels within the leach solution will markedly diminish with the precipitation of jarosite or hydroxides of iron. A further observation (still under investigation) is that the amount of ferric iron in the discharge solution tends to decrease although the leach rate increases. This effect may suggest that bacteria have a direct mechanism in the leaching of copper sulfides and that their role is not limited to the production of ferric iron.

4.4.3 Nutrition

Leaching bacteria require ammonium nitrogen and phosphate, supplied as $(NH_4)_2SO_4$ and either KH_2PO_4 or H_3PO_4. Dosages of these nutrients should be 10-20mg/L NH_4^+ and 30-40mg/L PO_4^{3-} for bioleaching. Care must be taken during addition of the ammonium since jarasite formation which results in the removal of iron from the leach cycle is favored and may cause heap permeability problems. To minimize these problems the heap should be at a pH < 2. Analysis of the ore and solutions is required to determine the level of nutrients and the presence of detrimental ions. Generally there is sufficient phosphate from mining activities and only ammonium addition may be required.

4.4.4 Heat

It is generally agreed that *T. ferrooxidans* and *T. thiooxidans* grow best at 20-35°C, although activity is evident outside these ranges. Temperature is important and a general rule of thumb is that bacterial activity halves for every 7°C temperature drop.

The heat within a leach pile is generally determined by the following major variables-local climate (ambient temperature, solar radiation, wind velocity), evaporation rates, heat of reaction, placement temperatures, irrigation temperature, irrigation strategy and irrigation rates. These variables must be addressed during design and operation of a leach pile. A favorable local climate may eliminate the effects of some of these variables, but all variables must be considered when operating in severe climatic regions. Heat modeling can help greatly in identifying the effects of variables and addressing them during design.

Evaporation is an important (sometimes forgotten) heat consideration. It is effected by factors including irrigation rates, irrigation strategy, heap and solution temperatures and local climatic conditions. Historically, copper operations have used whobler type irrigation systems, but high evaporation rates associated with this system result in high temperature losses. Several newer operations (Quebrada Blanca, Cerro Colorado and El Abra) use drip-type irrigation to avoid low temperatures in the heaps, even in warm climates. In the drip irrigation Quebrada Blanca Operation, waste heat from a power plant is added to the irrigation solution and the heap is covered with a permeable "shade" cloth to reduce the cooling effect of evaporation. In this operation the ore is heated prior to agglomeration and nearboiling water is added to assist in agglomeration and to maximize the agglomerate placement temperature. In this way a solution temperature of over 20°C is maintained throughout the year, even in a severe climate were the annual temperature averages only 5°C.

Production of heat due to the exothermic nature of the leaching reaction is only significant if high quantities of sulfides are oxidized over relatively short periods of time. Heat is an integral part of bacterial leaching and must be considered in any heap design.

4.4.5 Mineralogy

The mineralogy of the ore and gangue constituents affect bacterial activity as described above. Pilot testing is required on each ore to determine its suitability for bacterial leaching. The quantity and access of sulfides to bacterial leaching are both very important. The copper sulfide minerals are not alone in determining bacterial leach rates, but association with other minerals such as pyrite (which helps activity) or atacamite (which reduces activity) are important factors. This chapter is too short to review all aspects of mineralogy as related to bacterial leaching, however, involvement of a process mineralogist from project initiation is important.

4.4.6 Bacterial inoculation

In many new operations the inoculation of bacteria into the pile to help start or to sustain bacterial populations is considered. In other operations the right environment for bacterial growth is present and deliberate inoculation is not carried out. No reports on the benefits of bacteria inoculation in copper bioleaching have been found. Consideration must be given to the ability of the bacteria to adapt to the change from synthetic media to leach media. Bacteria in a leach pile also adapt and performance improves with time. Growth of bacteria in synthetic media does not allow for this adaptation process. An important variable when starting a new heap leach pile is to maintain a suitable acid level with very high-grade chemicals since the soluble copper sulfate produces the equivalent in acid during solvent extraction.

4.4.7 Iron

Bacterial leaching of copper is primarily concerned with the conversion of ferrous to ferric iron. There has been much discussion on the iron levels required for efficient copper leaching. It is easy to calculate the amount of iron required to leach a given quantity of copper if only a once-through ferrous to ferric conversion is assumed and if only ferric sulfate leaches the copper. What complicates this calculation is that ferrous to ferric iron conversion can occur multiple times within a pile depending on bacterial activity and the availability of oxygen. In addition, if copper conversion is also a direct result of bacterial activity, this calculation is unreliable. There are plants which operate successfully with iron levels in excess of 10g/L to levels below 2g/L. Iron is definitely necessary for the leaching reaction, but the absolute levels required are dependent on the factors that affect bacterial activity. Iron has been introduced in several start-up operations whereas natural ton build up has been allowed in others.

4.5 Heap Operating Variables

4.5.1 Irrigation distribution

Earlier reference was made to sprinkler (Whopler) and drip irrigation systems. Most copper bacterial heap leach operations have converted to drip-line irrigation in order to reduce the effects of evaporation. Various makes of drippers are used and most operations have ongoing programs to test different dripper types for drip-rate consistency and resistance to plugging and other failures. Drippers can cause a "dome" effect that may require moving the drippers from time to time. The drippers may be plowed into the pile in areas where freezing is common. Special equipment has been designed (Fig. 4.2) to

Fig. 4.2 Installation of drip lines above and below the heap surface at the Quebrada Blanca Operation, Chile

reduce labor for drip-line installation. Most operations maintain extensive systems for managing solution distribution, especially when hundreds of kilometers of drip lines spread over a large heap are involved, each at different point in the leach cycle (Fig. 4.3).

Fig. 4.3 Typical drip irrigation distribution system at the Quebrada Blanca Operation, Chile

4.5.2 Solution stacking

Solution stacking refers to the practice of recycling the pile effluent back to either the same pile or another pile at a different stage of leaching. This has been most commonly practiced in dump leach operations of low-grade ores to increase solution concentrations or to increase leach area without an increase in solution low. The use of solution stacking is accompanied by increased evaporation losses, pond usage and pumping requirements and a decrease in solution temperatures.

Solution stacking has generally been restricted to dump leach operations, although several heap leach operations now use this technique to improve solution discharge grades or to recycle higher grade ferric solutions to poorly leaching areas. The ability to recycle some solution is worth considering during plant design.

4.5.3 Solution collection

As described earlier, dump leach operations generally rely on the natural segregation of coarse and fine ore to provide natural drainage below the pile. Many of these piles were former waste dumps that are later leached, or are run-of-mine ore which is delivered to a prepared pad or a hillside. Environmental regulations usually require installation of membrane liner below the pad before ore placement. These pads usually have a minimum of drainage piping installation below the ore.

In heap leach operations, solution collection systems are vital to reduce the phreatic heads and stabilize stacked leach ore. A high density polyethylene (HDPE) liner is installed below the pad and then HDPE "Drainaflex" tubing is laid down in various configurations to collect the leach solution. The understanding of geotechnical issues has improved to such an extent that the phreatic head can be calculated during design and the correct drainage piping installed. During design of a solution collection system consideration must be given to the slopes of the pad, earthquake potential, material permeability, environmental protection and related factors.

The phreatic head should be maintained as low as economically possible in order to allow air penetration with good pile stability. Expense should not be the overriding consideration where drainage is concerned, as failure to collect the valuable product from a heap can be catastrophic.

4.5.4 Pad stacking/configuration

A wide variety of stacking equipment is available. The most common are the cascade or grasshopper conveyors, a continuous crawler type, or a stacker fed directly from a truck. Stacking of the ore for heap leaching is critical to maintain permeability in the heap. The cascade or truck type of stacker have generally been favored to reduce traffic on the heap (Fig. 4.4).

Fig. 4.4 Agglomerate stacking at the Quebrada Blanca Operation, Chile

There are basically two types of heap operation-the permanent heap and the ON/OFF or dynamic heap. In a permanent heap, fresh ore is stacked on top of leached ore, whereas in the ON/OFF system ore is removed from the prepared pad before placing fresh ore.

In the permanent pad it is usual to prepare the top surface of the pad before placing another level of fresh ore in order to provide a good surface for equipment, reduce solution inventory and provide a solution bleed to reduce build up of ions. Preparation of the various lifts is done either by rolling and compacting or by installation of a plastic liner between lifts. In some operations an interlift impermeable liner is not provided in order to allow for further recovery of copper from the lower lifts.

The ON/OFF pad offers the advantage of a reduced preparation area, but at the expense of removing the leach ore from the leach area. Another problem associated with the ON/OFF pad are the problems encountered when the leach time is longer than expected. Secondary leaching of removed ore can be introduced if suitable space is available.

4.6 Leach Solution Processing

The solution discharged from a leach pile (usually called Pregnant Leach Solution or PLS) requires further processing to achieve a marketable product. This PLS solution usually contains 1.5-6.0g/L copper, up to 20g/L ion and a variety of other ions (calcium, magnesium, sodium, manganese, potassium, chlorides, etc.)—all in a high sulfate solution. Another aspect which should be considered is that during processing, the soluble ions as well as the resulting acidity when copper is removed leave the sulfate to form sulfuric acid.

4.6.1 Copper cementation

The most traditional way of recovering copper from acid solutions has been by cementation with iron. The copper is allowed to make contact with iron filings or other fine iron products at a ratio of three tonnes of iron per tonne of copper to form a cemented copper precipitate.❹ The precipitate is washed, dried and sold to a refinery for further processing. The solution is recycled to leaching. This process is simple, but is affected negatively by the cost of iron, the low return on the sale of copper and the high manpower requirements. One of the few remaining operations using this method is the La Cascada Operation in northern Chile.

4.6.2 Direct electrowinning

An early alternative copper cementation, introduced by Inspiration Consolidated Copper Co. in 1929, was direct electrowinning (EW) of leach solutions. This method has also been practiced in Zaire and Zambia. The copper solution was fed directly to the electrowinning plant to produce an impure cathode containing copper, iron, zinc and other impurities. This product was sold as a subgrade metal.

Furthermore, direct electrowinning can be used to process the brass scrap. Comprising a concurrent electroleaching–electrodeposition process, it has been introduced as a potentially viable method for the selective recovery of copper. It has the advantages of process simplicity, low energy costs and much less environmental pollution as compared to the secondary smelting operations. M. Aghazadeh et al. recently studied a method for the direct recovery of copper from brass (Cu-30 wt.% Zn) scrap based on simultaneous electrolytic dissolution of the scrap at the anode and electrodeposition of copper at the cathode in an acidified sulfate electrolyte. The experimental results showed a smooth and compact copper deposit could be obtained at high Cu^{2+} concentrations and high current densities (above 40g/L and 250A/m^2, respectively) regardless of the level of other factors.

4.6.3 Solvent extraction

Copper solvent extraction (SX) in its modern form began in the mid 1960s when General Mills (now Henkel) produced its first selective solvent reagents. The first commercial operation was installed at the Bluebird Mine, Arizona, USA in 1968. The reagents have been under continuous development since that time. The newest reagents are highly selective against iron and can tolerate a much wider range of copper and acid conditions. These developments have led to the production of very pure copper solutions, with subsequent production of a highly marketable copper product from the electrowinning plants. This high quality copper cathode is usually sold directly for the production of copper bar or wire, brass, or other finished copper products without the need for additional refining. By the year 2000 it is expected that 20% of western copper production will be by use of the SX/EW process.

The basic copper solvent extraction circuit consists of three closed solution loops. In the first loop, the PLS solution from leaching is fed to the extraction portion of the SX circuit and contacted with an organic mixture of the selective extractant. The copper is extracted from the PLS solution to the organic solution. The two solutions being immiscible, separate and the barren solution (raffinate) is then reused for further leaching.

The second loop, within the SX plant proper, contacts the now copper laden (loaded organic) solution with a highly acid spent electrolyte to strip the copper from the organic. The now barren organic flows back around the loop to the extraction circuit.

Finally, in the third loop, the strong electrolyte is sent to the electrowinning plant. High purity copper is deposited by electrolysis and the spent electrolyte is returned back to the SX stripping section (Fig. 4.5 and Fig. 4.6).

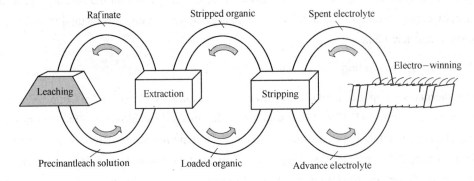

Fig. 4.5 Recycle loops in a Leach/SX/EW circuit

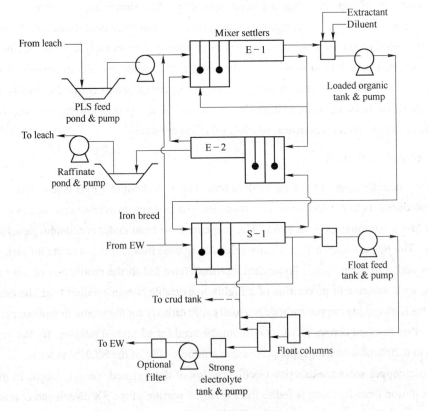

Fig. 4.6 Typical SX flowsheet

Solvent extraction provides several benefits. The acid is regenerated and only the acid consumed by the gangue is required for leaching. The organic is very selective, purifying the copper solution to EW to allow for production of a very high quality copper product. In the SX circuit there is some carryover of iron to EW. This iron is usually returned to SX using a small bleed stream of electrolyte. A problem is that fine solids from leaching can form a solids/organic crud that must be periodically removed from the SX circuit.

4.6.4 Electrowinning

Electrowinning is carried out in a number of acid resident cells through which electrolyte is circulated (Fig. 4.7). The cells contain a lead alloy anode and either a stainless steel cathode or a copper starter sheet cathode.[5] The cathodes with their copper deposit are typically harvested in a seven-day cycle. Copper is removed from the stainless steel blanks either manually or with an automatic or semiautomatic stripping machine. The final cathode product is typically one meter square and weighs 40-55kg/sheet. A small amount of cobalt sulfate is added to the electrolyte to help stabilize the anodes and a smoothing agent such as Gaur Floc is added to reduce cathode nodularization. Electrowinning is usually carried out at 40-45°C.

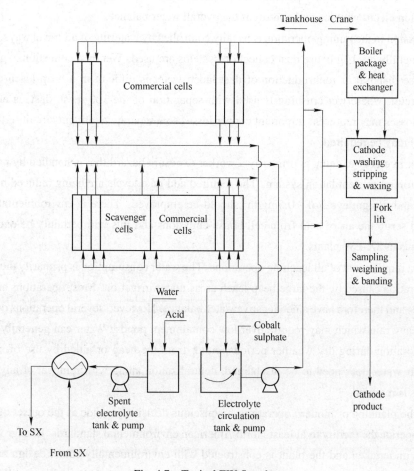

Fig. 4.7 Typical EW flowsheet

Most larger modern plants use polymer concrete EW cells with stainless steel permanent cathodes. The plant may also contain fully automatic cranes. The electrolyte is distributed within the cell through a manifold located at the bottom of the cell. An automatic stripping machine is used for cathode harvesting. The electrolyte is heated with a weak-to-strong electrolyte heat exchanger and also from an external boiler. During the plating cycle, oxygen is generated at the anodes which liberate some acid mist that must be controlled within the cell house. Acid mist can be controlled with a layer of balls, beads or a mixture of each placed on top of the cells. A plant may have a cross-flow forced air ventilation system. Some plants employ a surfactant (FC1100) to reduce the solution surface tension. Surfactants must be tested to determine any potential detrimental effects on SX operation or potential negative effects on leaching.

4.7 Commercial Installations and Environmental Considerations

A key feature of copper bioleaching, followed by solvent extraction and electrowinning is that it is an environmentally friendly process. It produces finished high quality copper without the need for a conventional concentrator or smelter. There are minimal process emissions or active tailings. All process solutions are recycled. The only environmental concerns are dust control in mining and crushing, acid mist control in electrowinning and control of the overall water balance.

Dust emission in the mining operation is usually controlled by watering of all travel ways. In crushing circuits, generally dry bag house dust collection systems are used. Wet dust collection is generally not used because disposal or reintroduction of dust laden water is difficult in a heap leach circuit. The Zaldivar circuit, where wet crushing is used with separation of the-100 mesh dust, is an exception. Also, dust losses may represent important copper losses so a separate treatment circuit (either leaching or flotation) may be required.

Acid mist in electrowinning is primarily a worker exposure hazard that is handled by passive ventilation or a forced air ventilation system. The required add mist levels are being reduced by regulatory authorities and design levels of <0.5mg/m^3 should be employed. There is environmental pressure to capture and scrub the air of acid from cell house emissions and this will certainly be one of the next developments in SX-EW plants.

Water is a requirement of all leaching operations. The water consumption is primarily the result of the high evaporation caused by the large heap leach areas under irrigation. Most operations are located in dry climates and therefore have a net negative water balance. However, several operations do experience periodic heavy rain which may require overflow containment ponds. Water can generally be disposed of by evaporation during dry weather periods, either from the heap or aided by use of sprinklers on the heap. If water does need to be discharged, neutralization along with removal of any SX organic compounds is required.

During the planning of modern operations a conscious decision is made at the outset of a project to build and operate the facility to at least North American environmental standards. Background baseline studies are undertaken and the plant is constructed with environmentally sound design and operating features. The leach pads use welded HDPE liners, areas are contoured so that spillage will be contained and ponds and tanks have emergency containment facilities. Extensive environmental monitoring is part of normal operations.

This section presents three examples of operations which demonstrate the three main types of bioleaching used: in situ leaching, dump leaching and heap leaching. Little work has been carried out in the bioleaching aspects of in situ and dump leaching, although bacteria are known to be present. Examples given are intended provide an overview of leaching technology as applied commercially to the production of copper.

4.7.1 In situ leaching

In situ leaching represents leaching within an undisturbed ore body where no mining activity is required. There is limited evidence that bacteria play a role in situ mining. On the other hand, existing secondary deposits are the result of bacterial action over an extended period of time. In the interest of completeness, the in situ operation at the Sari Manuel operation of BHP Copper is presented to illustrate most of the principles of in situ leaching.

Sari Manuel is located in the southwestern United States about 60km northeast of Tucson, Arizona. Underground mining of sulfide ore since 1955 has resulted in substantial surface subsidence. In 1985 Magma Copper began an open pit operation to recover oxide ore by acid leaching, followed by an SX/EW circuit to retrieve the copper from the leach solution. Owing to pit geometry, mining economics and the irregular distribution of the remaining open pit ore, an in situ alternative was considered to recover the remaining ore using the existing SX/EW facilities.

Initial in situ mining began in 1988 using an array of wells to inject acidified leach solution into the mineralized ground. The leaching solution was gravity driven through the ore, collected in the abandoned underground workings and pumped 725m to the surface. It was difficult to coordinate injection sites because these depended on open-mining activity, fluid how paths and underground collection areas which were already in place. Further development was undertaken to improve control of this in situ leach operation. A subsequent well-to-well in situ leaching operation that used pumping wells to collect injected leach fluids was implemented in 1989. After open-pit mining ceased in 1995, production continued at a rate of 20,000t of copper cathodes per year as a result of this in situ leach program.

Drilling was undertaken to define the distribution of copper grade. These data were used to define zones favorable for in situ leaching and to design the pumping system. The basic well pattern employed is a seven-spot cell consisting of six injection wells distributed around a central production well. Individual cells are linked by shared injection wells along the central axis of a linear well pattern. The depth rid screening of wells is based on subsurface grade and structure. The wells were drilled by reverse circulation with a 3.8 on casing diameter and the production wells are 15.25 on in diameter. Wells were drilled to 100-150m depth and cased with PVC. Fluid how rates in injection and production wells are logged hourly and the system shows a 13.5% fluid loss.

The leach solutions are a combination of raffinates from the ongoing heap operation and the in situ operation. During just over a year the acid content of the injection fluid was maintained at 26g/L. Copper concentration in the production fluid started at about 2g/L and then plateaued at 1.1g/L for a leach time of over two years. Overall recovery is almost negligible at starting grades of <0.2% copper, but rises to above 80% for grades of over 1.5% copper. Overall recovery of a test section averages about 60%. This depends also on the ratio of oxide to sulfide minerals present. A significant effect is formation of gypsum and jarosites which is caused by the mineralization of the gangue materials combined with the high dissolved solids content of the leach solution.

This in situ operation is producing cathode copper from a low-grade irregular ore body that would have become waste. The low operating costs combined with an existing facility continue to provide an economic process. The future will allow for development of in situ leaching for new low-grade deposits with minimal disturbance of the surrounding areas.

4.7.2 Dump leaching

Most traditional dump leach operations have involved the leaching of existing waste piles to extract low-grade ore that was previously considered waste. Several operations for leaching low-grade run-of-mine ore dumps have been designed. A good example of an operation that has a "designed" dump leach rather than an "after the fact" waste dump, is the Baja Leg operation of the Chuquicamata Division of Codelco, Chile.

The first studies for this facility started in 1970, with conceptual and basic engineering design commencing in 1983. Construction began in 1991 with plant startup in 1993. The plant was designed for 15,000 tonnes of cathode production from a 0.35% copper run-of-mine Anger dump operation at a cost of U.S. $40 million.

The ore is placed with mine trucks in 35m high 'fingers' 150m wide. Leach irrigation modules of 70m×35m, using drip irrigation are established on top of these Angers. The leach cycle consists of preconditioning, leaching, rest, conditioning, and wash cycles over a 78-week period. The solution used 1s the raffinate from the adjoining SX/EW plant. Acid consumption is 9.5kg/kg of copper produced. The resulting PLS solution is collected at the bottom of a valley in a 34,000m^3 PLS pond. This solution is then transferred to a modern SX/EW plant employing stainless steel cathodes and a semiautomatic stripping machine.

The plant employs 27 workers and produces copper at less than U.S. $0.40 per pound. Overall recovery is estimated at 20%. This plant is an example of a successful design run-of-mine dump leach operation on a low-grade ore, producing finished copper at low a cost.

4.7.3 Heap leaching

The Quebrada Blanca operation is located in northern Chile in the Alti Plano desert at 4,400m above sea level. It is a stand-alone crushing, bacterial leach, solvent extraction and electrowinning facility. The operation includes its own power plant and obtains water from a Safar 38km from the mine. This facility processes 17,300t/d of sulfide ore to produce 206t of London Metal Exchange (LME) grade cathode copper per day.

The property was explored in the late 1970s, drilled and an underground exploration adit was developed which provided valuable samples for pilot testing. Conventional processing was considered to be uneconomical. In 1988 the property as offered in an international tender and pilot studies were initiated with SMP Tecnologla S.A. After feasibility studies, a decision to proceed with production was made late in 1991. Plant start-up was in January 1994, with the first cathode produced in August 1994. The Quebrada Blanca operation is based on a supergene enriched porphyry copper (primarily chalcocite) deposit with a reserve of 91Mt at 1.3% copper.

Sulfide ore from the mine is crushed in three stages to 100% minus 9 mm. The crushed ore is fed to a 1,500t surge bin that includes a steam-to-ore heat exchanger to increase the temperature of the ore. The ore is then agglomerated in rotating drums with 5-7kg/t of sulfuric acid and 85°C hot water to obtain about 10% moisture in the final agglomerate. The ore is discharged from the agglomeration drums at 22-24°C and is conveyed to the stacking circuit. The agglomerate is moved from a tripper conveyor to a series of shiftable conveyors which transport the ore to a shuttle conveyor and stacker. The stacker distributes the ore uniformly in an arc to form a 6-6.5m high continuous pile or sector, 85m in width.

The leach pile is lined with 60mm HDPE welded liner. A network of 4" dia. Drainaflex pipe is placed at 2m intervals on the liner or on compacted leached ore (after the bottom lift is completed). Immediately after the ore is stacked, two networks of drip irrigation lines are installed on top of the pile, one on the surface and the other 20cm below surface. The leach area is irrigated at 0.1-$0.14L/(min \cdot m^2)$ of 7.0g/L sulfuric acid raffinate solution from solvent extraction. The raffinate solution is heated to 28°C in a set of heat exchangers prior to being fed to the drip lines. Heating of the raffinate solution also serves to cool the generators in the power plant.

The 3.5s/L Cu PLS solution at 23°C is collected in the Drainaflex lines and flows to a series of collection header lines that are buried within the leach pile. The leach circuit also includes several innovations to help improve leaching under the severe weather conditions found at this site. A separate series of air lines is installed below the heap to distribute air from a series of air fans. Nutrients are added to the leach solutions to maintain adequate ammonia and phosphate levels for bacterial activity. The top of the pile is covered with shade cloth to reduce evaporative cooling. This operation carries out extensive heat, oxygen, solids and liquid monitor rig to help optimize leaching. This monitoring includes on-site bacterial activity measurements. Measurements from residue samples indicate that a recovery of better 80% of total copper is obtained.

Over 3,000 m^3/h of PLS solution at over 20°C and at 3.0-3.5g/L copper is for warded to the solvent extraction facility which consists of three process trains in parallel, each with two extraction stages followed by one stripping stage. The PLS is contacted in a counter-current stream of organic solvent to extract the copper in two stages with a recovery efficiency of 93%. The solvent consists of up to 13.5% LIX 984 carried in a low aromatic kerosene (SX_{12}). The loaded organic solvent reports to a loaded solvent surge tank, while the aqueous raffinate is returned to aching. The loaded solvent is advanced to the stripping stage where it is contacted with 35g/L Cu lean electrolyte in two mixer stages to transfer the copper to the aqueous electrolyte solution. The barren organic solvent is advanced from stripping to extraction, while the 49g/L Cu rich electrolyte is discharged from the strip stage to feed two flotation columns in series, followed by a garnet/anthracite pressure filter for removal of entrained organic solvent.

The EW cell house contains 264 polymer concrete cells, each with 60 permanent stainless steel cathodes and 61 lead/tin/calcium alloy anodes. The cell house is operated at a nominal 260A/m² for a seven-day cathode growth cycle. One-sixth of the cathodes are harvested daily using an automatic WENMEC stripping machine, and the stripped blanks are returned to the cells. A bottom wax is applied to each cathode. Guar is added to reduce nodule formation and 100ppm of cobalt sulfate is used to reduce anode corrosion. All water used in the cell house is used in a reverse osmosis plant to reduce chloride levels. The stripped LME grade cathodes, 45-50kg each, are stacked in 2.5t bundles and strapped for truck haulage to Iquique for shipping to the final destinations.

This operation produces high quality LME Grade A copper at around U.S. $0.50 per pound from an initial capital investment of U.S. $360 million. It demonstrates an advancement of the application bioleaching of copper at a large scale to a region with adverse climatic conditions.

4.8 Summary

This chapter has reviewed the many elements that must be considered in planning, designing and operating a successful copper bioleaching operation. Outwardly, this technology appears very simple, but there are many factors that must be controlled to keep the bacteria alive and active. Existing operations and research organizations are continuing to advance the knowledge base required to make this technology more reliable. The few operations included in this chapter demonstrate that ore previously considered to be waste can indeed be mined economically using bioleaching technology.

The future will see this technology expand to include more difficult-to-leach ores such as chalcopyrite and the treatment of copper concentrates. Bioleaching is a low cost technology that is environmentally friendly and it produces a high quality finished product. It is an alternative that must be considered whenever a new ore body is considered for development.

Questions

1. What are the common copper minerals considered in heap bioleaching?
2. What are common variables which can affect leaching rates and recoveries to heap, dump, or in situ leaching systems?
3. What are the variables of bacterial leaching of copper?
4. How to leach the solution with the metal ions?
5. How does the copper solvent extraction work?

Note

注　释

❶ 原位浸矿是不移动矿体，采用矿坑或钻孔直接在原地位置进行浸矿的方法。

❷ 在铜的溶解过程中，铁离子是主要氧化剂，生物浸矿过程产生铁离子是其优势之一，可取代生物浸矿的有效方法就是直接添加铁离子。

❸ 铜的化学浸矿初期受酸的添加量影响，但后期为保证细菌的正常生长，浸矿过程必须保证稳定合理的 pH 值。

❹ 从酸溶液中回收铜，最经典的方法是采用铁离子凝合法。铜可以被铁粉或其他的细铁产品结合，比率是每三吨铁结合一吨铜，然后凝合。

❺ 电解提取法利用系列电解槽循环使用电解质，电解槽中阳极是铅合金，阴极是不锈钢或者铜始极片。

5 Biooxidation of Gold-Bearing Ores

本章要点 在生物提取黄金中，各种因素会影响细菌群组成，如高温、低 pH 值会使螺旋菌数量增加，所以使用混合菌时要首先确定菌群性质。溶液中较高浓度的亚铁离子可促进液相里细菌群的增加，但较高浓度的亚铁离子会降低氧化还原电位从而改变反应过程。硫化物氧化是放热反应，而嗜温菌氧化反应的合适温度为 30~45°C，为提高硫化物矿物氧化率，有必要对生物反应器降温并保证其温度范围。生物氧化过程的效率依赖于细菌存活状态，某些化学试剂或天然元素对细菌来说是有害的。

5.1 Introduction

In 1984, a pilot plant was commissioned at GENCOR Process Research to treat Fairview flotation concentrate and was operated for some two years. Its success led to construct an industrial plant at Fairview to treat 40% of the mine's production. In 1991, the Fairview BIOX® plant was extended to treat the total production of the mine, amounting to 40t/d of flotation concentrate, containing up to 160g/t of gold.

The BIOX® process has been both a technical and economic success and is fast becoming the norm in refractory gold ore processing. Biooxidation offers real economic advantages over roasting and pressure oxidation, and the BIOX® process has been clearly demonstrated to be environmentally acceptable and robust enough to be operated in remote areas.

Within the GENCOR group, a second plant was built in 1986 at Sao Bento Mineracao in Brazil, and in 1988 the company took the unusual step (for a South African mining house) of entering the international technology licensing business. This decision formed part of GENCOR's thrust to establish itself more strongly internationally.

The first licensed plant was commissioned at the Harbour Lights mine, Western Australia in 1992. Asarco Australia also decided to use the BIOX® process at its Wiluna operation in Western Australia. This plant, with a capacity of 115t of concentrate per day, was successfully commissioned in 1993. The plant was to be expanded in 1996 to treat 155t/d.

The BIOX® process took a quantum leap with the construction of a 720t per day BIOX® plant at the Ashanti Goldfields Company's Obuasi mine, at Obuasi in Ghana. The plant design consisted of three independent BIOX® modules, and construction was completed in January 1994.

Laboratory and pilot scale testwork to evaluate the amenability of ores to biooxidation remains a major activity at GENCOR Process Research. To date, over 150 ore samples have been subjected to amenability tests at laboratory scale and over 12 integrated pilot runs have been completed, to provide information for commercial plant designs and project feasibility studies.

Process development is ongoing in order to improve the performance of biooxidation with respect to reducing operating costs and improving process performance and operability. A major development in recent years has been the use of the LIGHTNIN A_{315} fluid foil impeller instead of the conventional radial flow turbines for gas dispersion and solid mixing, resulting in a 30% power saving for the operation of large scale bioreactors. The development of rate models describing biooxidation and models describing gas liquid mass transfer in bioreactors has been undertaken, to allow improved optimization of the process design specification.

5.2 BIOX® Bacterial Culture

The heart of the BIOX® process is the bacteria; they are a mixed population of *Thiobacillus ferrooxidans*, *Thiobacillus thiooxidans* and *Leptospirillum ferrooxidans*. They are all chemolithotrophic Gram-negative acidophiles and are motile by means of a single polar flagellum. *T. ferrooxidans* and *L. ferrooxidans* oxidize reduced iron compounds for energy, whereas *T. thiooxidans* obtains energy from reduced sulfur compounds, as can *T. ferrooxidans* also. The thiobacilli are straight rods with a diameter of 0.3-0.6μm and a length of 1-3.5μm. Although *Leptospirillum* has similar dimensions, it is vibroid when young and spiral when older. It is also highly motile. *Leptospirillum* is not related to the thiobacilli and neither are *T. ferrooxidans* and *T. thiooxidans* related to each other as indicated by DNA-DNA similarity values.

The bacteria collaborate in oxidizing metal sulfides such as arsenopyrite and pyrite and facilitate this by becoming attached to the ore. When bacterial attachment was examined by the DAPI staining technique, it was found that they tend to attach specifically to the metal sulfide. This has been confirmed by other investigations. There has been disagreement about the relative importance, and mechanism of direct and indirect leaching, where indirect leaching involves the production of ferric sulfate by the bacteria, which causes chemical breakdown of the metal sulfide. Attachment tends to support the idea of direct bacterial involvement. Recently, it has been claimed that all leaching is indirect. The specific rate of oxygen utilization in the presence and absence of pyrite was found to be the same and to follow the same kinetics. From this it was concluded that pyrite is oxidized by means of an indirect mechanism, in which it is leached chemically by ferric iron and the role of the bacteria is to generate ferric iron and maintain a high redox potential in the system.

These conclusions were based on an investigation where the bacteria consisted mainly of *L. ferrooxidans* with less then 5% being *T. ferrooxidans*. Since *Leptospirillum* can only oxidize ferrous iron, it is likely that pyrite breakdown by this bacterium does occur by the indirect leaching mechanism. However, this is not necessarily true for *T. ferrooxidans* which can also oxidize sulfides.

Since the composition of the bacterial population can be influenced by various factors such as high temperatures and low pH which enhance *Leptospirillum* numbers, it is important to determine the nature of the mixed population.❶ This, in turn, can have a profound effect on the quality of the leach. The dot-blot immuno binding assay as well as the PCR technique have been used to analyze bacterial populations.

The BIOX® population has been investigated with the microscopic immunofluorescence technique. This was made quantitative by counting the total number of bacteria in the same field of view using phase contrast after the fluorescent bacteria (which may be *Leptospirillum*, *T. ferrooxidans* or *T. thiooxidans* depending on the fluorescent antibodies used) had been counted under UV light. A number of microscopic fields were counted for any one species.

The Sao Bento BIOX® tank was found to contain 48% *L. ferrooxidans*, 34% *T. thiooxidans* and 10% *T. ferrooxidans*, as free unattached bacteria. A similar proportion of bacteria was determined in the Fairview multistage continuous plant where in the primary tank (equivalent to the Sao Bento tank) *Leptospirillum* made up 57%, *T. thiooxidans* 30% and *T. ferrooxidans* 13% of the population of free bacteria. The composition of the bacteria attached to the ore was obtained by releasing them into acid water (pH 1.8) using Triton X_{100} and vortexing. Of this, 57% was again *Leptospirillum*, with 26% *T. thiooxidans* and 17% *T. ferrooxidans*. Since, counting errors do arise, it is possible that no real difference exists in the proportions of free bacteria and attached bacteria.

However, in progression towards the final tank, the proportion of *T. thiooxidans*, particularly those attached to the ore, was found to increase while less *Leptospirillum* remained attached. This can be explained by a decrease in oxidizable iron and an increase in sulfur intermediates which provide an energy source for *T. thiooxidans*. Over the cascade of tanks, the amount of free *T. ferrooxidans* dropped to a low 5%, while the amount attached increased slightly and then remained relatively constant. That *L. ferrooxidans* plays a more dominant role in the continuous process than *T. ferrooxidans* has been confirmed by other studies. It has also been indicated that a predominance of *Leptospirillum*-like bacteria increases the iron-leaching rate by a factor of approximately 2.4.

In ores rich in arsenopyrite, as is the case for many gold-bearing refractory ores, the first step in leaching is growth of bacteria attached to the ore. This enhances direct bacterial oxidation, causing the appearance of the first corrosion patterns on the mineral and solubilization of Fe^{2+}, As^{3+} and S^{2-}. Subsequently, the presence of high concentrations of Fe^{2+} iron in solution enhances the growth of free bacteria and Fe^{2+} and As^{3+} become oxidized to Fe^{3+} and As^{5+}. The Fe^{3+} iron generated plays a role in arsenic and sulfide oxidation and in precipitation of ferric arsenate, thereby coating the attached bacteria.

Bacteria growing on arsenopyrite may be inhibited by the arsenic released. The BIOX® bacteria are tolerant to arsenic (V) concentrations of 15-20g/L. However, they are less tolerant to arsenic (III) and become inhibited above arsenic (III) concentrations of 6g/L. Just as arsenic can be toxic to the bacteria, so can other constituents of the ores, such as antimony and mercury. Tolerance limits of the BIOX® bacteria towards these two metals is 10% by mass for antimony and at least 0.12% for mercury. An unusual feature of the acidophiles is their high sensitivity toward chloride ions, the reason for chloride toxicity appears to be membrane damage.

Other environmental factors that impact on the BIOX® bacteria are pH, temperature, carbon dioxide and nutrients. When shake-flask tests were carried out on BIOX® *T. ferrooxidans* inoculated into 9K medium with pH adjusted to 1, 1.6, 1.8, 2.3 and 4, it was found that oxidative activity was inhibited at a pH less than 1.6. The optimum was in the range of pH 2-3. Both *L. ferrooxidans* and *T. thiooxidans* were still active at pH 1, although *T. thiooxidans*, grown in a sulfur medium, showed little growth between pH

0.5-1.0. Thus, bacterial leaches at very low pH would have few *T. ferrooxidans* and more of the other bacteria.

Using a Chemap fermentor and varying temperatures between 25°C and 50°C with other parameters constant, the optimum operating temperature of the BIOX® bacteria found to be 40°C, although their oxidative potential was only slightly lower at 35°C and 45°C. The bacteria were not killed at 50°C, but their oxidation rate slowed down considerably, with the time required for complete conversion of ferrous to ferric iron increasing from approximately one day at 40°C to 3 weeks or more at 50°C. Growth of *Leptospirillum* will be favored compared to *T. ferrooxidans* in an environment with a low pH (<1.5) and a temperature of 40°C or higher. *T. ferrooxidans* will often be prevalent in batch cultures operating at high ferrous concentrations, and controlled pH and temperature of pH 1.4-1.6 and 35-40°C, respectively.

The effect of added carbon dioxide was assessed by the rate of iron oxidation using air (0.03% CO_2) and air supplemented with CO_2 to give concentrations of 0.3% and 1%. Tests were carried out in a Chemap fermentor at two different temperatures, 30°C and 40°C. It was found that the different concentrations of CO_2 had no effect at 30°C, while at 40°C added CO_2 markedly shortened the time for complete oxidation of ferrous iron. At 30°C, temperature was probably more limiting on oxidative activity than was carbon dioxide concentration.

The nutrient formulation used to grow *T. ferrooxidans* and *Leptospirillum* in the absence of metal sulfides is 9K. This contains basic salts and 50g/L ferrous sulfate. In the presence of metal sulfides, 0K, which lacks iron but is otherwise identical, is supplied. This formulation has been described as providing an excess if nutrients. Our laboratory investigations proved that this is the case and a modified 9K medium containing 50g/L $FeSO_4 \cdot 7H_2O$, 1g/L $(NH_4)_2SO_4$, 0.4g/L K_2HPO_4 and 0.1g/L $MgSO_4 \cdot 7H_2O$ did not diminish bacterial oxidation rates.

5.3 Chemical Reactions and Process Control

5.3.1 Chemical reactions

BIOX® process comprises contacting the refractory sulfide ore or concentrate with a preparation of the GENCOR BIOX® mixed bacterial culture for a suit able treatment period while maintaining an optimum operating environment. The bacteria cause accelerated oxidation of the sulfide minerals thereby liberating the occluded gold for subsequent recovery via cyanidation. The principal sulfide minerals associated with refractory hydrothermal gold ores are arsenopyrite, pyrite and pyrrhotite.

The bacterial oxidation reactions of the sulfide minerals pyrite, arsenopyrite and pyrrhotite, to achieve gold liberation, may be summarized as follows:

$$4FeS_2 + 15O_2 + 2H_2O \longrightarrow 2Fe_2(SO_4)_3 + 2H_2SO_4 \quad (5.1)$$

$$2FeAsS + 7O_2 + H_2SO_4 + 2H_2O \longrightarrow 2H_3AsO_4 + Fe_2(SO_4)_3 \quad (5.2)$$

$$4FeS + 9O_2 + 2H_2SO_4 \longrightarrow 2Fe_2(SO_4)_3 + 2H_2O \quad (5.3)$$

These reactions clearly indicate the high oxygen demand of sulfide oxidation. Also evident is the acid-consuming nature of pyrrhotite and arsenopyrite oxidation with pyrite oxidation being the sole acid-producing reaction. The oxidation reactions are also highly exothermic. In addition to the direct oxidation of sulfide minerals, several indirect chemical and bacterially assisted reactions occur as follows:

$$FeS + Fe_2(SO_4)_3 \longrightarrow 3FeSO_4 + S^0 \tag{5.4}$$

$$FeS_2 + Fe_2(SO_4)_3 \longrightarrow 3FeSO_4 + 2S^0 \tag{5.5}$$

$$2FeS + 2H_2SO_4 + O_2 \longrightarrow 2FeSO_4 + 2S^0 + 2H_2O \tag{5.6}$$

$$4FeSO_4 + 2H_2SO_4 + O_2 \longrightarrow 2Fe_2(SO_4)_3 + 2H_2O \tag{5.7}$$

$$2S^0 + 3O_2 + 2H_2O \longrightarrow 2H_2SO_4 \tag{5.8}$$

Important secondary reactions include precipitation of ferric arsenate ($FeAsO_4$), acid dissolution of carbonates and precipitation of jarosite, according to the following reactions:

$$2H_3AsO_4 + Fe_2(SO_4)_3 \longrightarrow 2FeAsO_4 + 3H_2SO_4 \tag{5.9}$$

$$CaMg(CO_3)_2 + 2H_2SO_4 \longrightarrow CaSO_4 + MgSO_4 + 2CO_2 + 2H_2O \tag{5.10}$$

$$3Fe_2(SO_4)_3 + 12H_2O + M_2SO_4 \longrightarrow 2MFe_3(SO_4)_2(OH)_6 + 6H_2SO_4 \quad (M^+ = K^+, Na^+, NH_4^+, H_3O^+) \tag{5.11}$$

5.3.2 Influence of ore mineralogy

The relative proportions of each mineral dictate various process requirements such as cooling, acid consumption/production, oxygen demand, degree of precipitation and neutralization. The oxidation and chemical leaching of principal refractory gold ore minerals have several process data, such as the heat of reaction, acid demand and oxygen demand. The actual values, for treatment of a particular concentrate, will be dictated by the relative proportions of the major minerals. Typically the overall heat of reaction is about 30MJ/kg sulfide with an oxygen demand of 2.2kg/kg sulfide oxidized. Examples often effects of the major minerals upon the operation of biooxidation and the BIOX® process follow:

(1) Pyrite. Bacterial oxidation of pyrite is highly acid-producing. Therefore, treatment of a concentrate with a high pyrite content will be acid-generating and maintenance of the pH within the required operating range requires addition of lime or limestone.

(2) Pyrrhotite/Pyrite. Due to the acid-consuming nature of pyrrhotite, the relative proportion of pyrite to pyrrhotite is an important factor affecting the overall lime and/or acid requirements, and one which also influences solution redox potential. The acid dissolution of pyrrhotite releases ferrous iron and elemental sulfur. Although the formation of elemental sulfur by this means is reversed by the bacteria present in the culture, excessive elemental sulfur formation, due to an abnormally high pyrrhotite content, cannot be accommodated in the plant and may lead to an increase in cyanide requirements and lower gold recovery.

The higher ferrous level in solution is beneficial in that it promotes a large bacterial population in the liquor phase of the primary reactors, which in turn reduces the possibility of bacterial washout occurring. However, the higher ferrous concentration lowers the redox potential, which can alter the oxidation chemistry of the process.❷ The most serious effect of low redox potentials, combined with a low iron to arsenic ratio in solution, is the possibility of arsenic (III) formation. Arsenic (III) may precipitate as the less stable calcium arsenite compound; hence formation of arsenic (III) should be minimized as far as possible. In addition, arsenic (III) has a greater toxicity effect upon the bacteria than arsenic (V).

(3) Arsenopyrite. The ratio of arsenopyrite to pyrite also influences acid consumption, but to a lesser extent than that of the pyrrhotite to pyrite ratio. More critical is the ratio of arsenopyrite to pyrrhotite and pyrite as given by the iron to arsenic ratio.

The iron to arsenic ratio is critical as it dictates the stability of ferric arsenate precipitates formed on neutralization of the BIOX® waste liquor. The molar ratio of iron to arsenic in a concentrate is generally required to be greater than 3 to achieve stable effluent products, with respect to arsenic solubilization.

(4) Carbonate minerals. Carbonate content has two major effects on the BIOX® operation. Firstly, a minimum content is required to ensure production of sufficient CO_2 to promote, bacterial cell production. If no carbonate is present, limestone must be added to the primary vessels, or the carbon dioxide content of the air injected must be further enriched with carbon dioxide.

The second effect is that of carbonate dissolution on pH. At a low sulfide to carbonate ratio, the primary stage becomes add-consuming. The degree of precipitation increases and results in coating of the sulfide surfaces. Formation of coatings may result in lower oxidation rates, which in turn reduces liberation of gold for dissolution on cyanidation. Presence of carbonate at a high sulfide to carbonate ratio is beneficial, not only for CO_2 production, but also in reducing lime requirements for pH control during biooxidation.

5.3.3 Effect of temperature and cooling requirements

5.3.3.1 Extent of sulfide oxidation

The BIOX® bacterial culture is an adapted mixed culture of mesophilic bacteria, as described earlier in this chapter. Continuous pilot testwork has shown that although this culture operates best at 40°C, operation at a temperature of 45°C is possible in the primary reactors (where initialbacterial growth occurs) and at 50°C in the final secondary reactors.

The effect of operating at 45°C for biooxidation of the Fairview and an Australian concentrate is shown in Fig. 5.1. It is noted that no significant decrease in the final sulfide oxidation occurs, compared

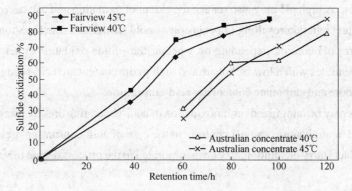

Fig. 5.1 Effect of temperature on sulfide mineral oxidation

to operation at an "optimum" temperature of 40°C.

5.3.3.2 Cooling requirements

The oxidation of sulfide minerals is extremely exothermic. Biooxidation with mesophiles requires operation at a temperature in the range of 30-45°C. At a high rate of sulfide mineral oxidation, cooling of the bioreactors is necessary in order to stay within this temperature range.[3] The possible heat sinks heat loads for biooxidation, excluding the relatively small radiative and convective heat losses, are as follows:

(1) Heat of reaction of sulfide mineral oxidation.

(2) Heat generation by absorption of agitation power.

(3) Sensible heat loss or gain from inlet air on adjusting to slurry temperature.

(4) Heat loss to heating the incoming slurry to the vessel operating temperature.

(5) Evaporative cooling provided by the sparged air at slurry temperature.

(6) Heat loss due to air expansion.

Typical heat loads for biooxidation plants are 30MJ/kg sulfide oxidized. The most cost-effective method for cooling is by circulation of cooling water through internal coils in the reactor and removal of the heat from the water in an evaporative cooling tower. The efficiency of evaporative cooling is dependent on the local climatic conditions, principally the wet bulb temperature. The efficiency is poor in equatorial regions with high ambient temperatures and high humidity.

Operating biooxidation at a temperature of 45°C compared to 40°C may reduce the cooling water flow rate required by 37% and the cooling tower cross sectional area by 25%. However, although this saving is significant it would only represent a reduction of about 5% in the total capital cost of a biooxidation plant and a negligible savings in operating costs. Nevertheless, operating biooxidation at 40°C gives a safety margin whereby the slurry temperature may increase by 5°C or more before bacterial activity is adversely affected. A temperature rise may occur on operating plants due to failure of cooling water supply, or scaling of the internal cooling coils if correct maintenance is not carried out.

5.3.4 pH control

Optimum rates of sulfide oxidation and gold recovery will require control of the slurry pH to maintain an active bacterial culture. A typical operating pH is in the range pH 1.2-2.0. A low pH reduces the extent of sulfide oxidation. A high pH may also reduce the extent of oxidation and can decrease gold recovery due to metal salt precipitation resulting in occlusion of gold particles. Lime/limestone or acid addition will be required for pH control, depending on whether the sulfide oxidation is net acid-producing or consuming. Concentrates with a low carbonate and high pyrite content are acid-producing, concentrates with a high pyrrhotite and carbonate content are acid-consuming.

Operating costs may be minimized for biooxidation of acid-generating ores by using a locally sourced limestone for acid neutralization and pH control. In the case of acid-consuming ores, recycle of acidic biooxidation product slurry or solution may be incorporated in the process design to reduce consumption of fresh acid.

In the case of treatment of refractory gold ore concentrates, upstream flotation offers the opportunity to produce a concentrate where acid-consuming minerals such as carbonates or pyrrhotite, are rejected to the flotation tailings. In designing a biooxidation plant, considerable attention must be given to the upstream concentration or preparation process to ensure that the optimum grade and grind of ore (or concentrate) is produced for treatment by biooxidation in order to ensure maximum gold recovery and thus, financial return.

5.3.5 Oxygen supply

The supply of air to biooxidation represents the largest consumer of power in the process and hence the major contributor to the overall operating cost. Typically, sulfide mineral oxidation requires about 2.2kg oxygen per kg sulfide oxidized.

For a biooxidation plant treating 240t/d of concentrate, at 12% sulfide and 90% sulfide oxidation, the sulfide oxidation duty is 1,080kg/h. The equivalent oxygen demand is 2,376kg/h or 7,912m^3/h of air at 100% oxygen utilization. In practice, the oxygen utilization will be much lower than 100%. Assuming a utilization of 30%, the air demand for biooxidation increases to 26,373m^3/h.

In order to attain acceptable oxygen transfer rates and oxygen utilization in the bioreactor, the air supplied must be well dispersed. Air dispersion is achieved by mechanical agitation. The power required for supplying oxygen, air delivery and dispersion, represents 30%-40% of the total power requirements for biooxidation of metal sulfide concentrates. The power for air supply can be minimized by designing reactors with a low height-to-diameter aspect ratio, so that the compressor or blower pressure head required is as low as possible. A low pressure head permits the use of blowers instead of compressors, which consume less power per cubic meter of air delivered.

The criteria for the agitator specification can be summarized as follows:

(1) Sufficient power must be provided to the impeller to prevent flooding; flooding occurs when air passes through the impeller and is not dispersed by fluid flow.

(2) The oxygen mass transfer coefficient ($kl.a$) for the agitator/aeration system must meet or exceed the required $kl.a$, determined by the oxygen demand of the process.

(3) The impeller pumping rate must be sufficient to achieve uniform solids suspension.

The air supply must be adequate to meet the process oxygen demand determined by the rate of sulfide oxidation and to maintain a dissolved oxygen concentration in solution of not less than 1.5 ppm in order to allow an adequate oxygen supply to the bacteria.

The input power necessary to attain the required oxygen uptake rates, for biooxidation of metal sulfide concentrates, is normally sufficient to be the controlling factor in the agitator specification. Radial flow impellers such as the Rushton turbine are the traditional impellers used for high gas dispersion rates. However, recently developed axial flow fluidfoil impellers, such as the LIGHTNIN A$_{315}$ impeller, offer improved efficiency, giving equivalent oxygen transfer rates at reduced power consumption. The fluid flow developed by these impellers is also much higher than radial flow turbines, per unit power input, allowing solids suspension at reduced shear rates and lower power levels.

The design of agitators for bioreactors has been well described by Fraser. The agitator must be selected to match the required reactor duty so that the overall power consumption for oxygen supply is minimized.

The solids content of the process slurry has a direct influence on the oxygen mass transfer rate, as shown in Fig. 5.2. Bailey and Hansford have demonstrated that oxygen mass transfer is the limiting factor controlling the solids content of process slurry in biooxidation. For typical sulfide concentrates at 20%-30% sulfide, the required oxygen transfer rates limit the solids content and hence sulfide content per unit volume, to about 20% by mass. For low-grade concentrates or ores at 2%-5% sulfide, solids concentrations of over 30% by mass may be tolerated.

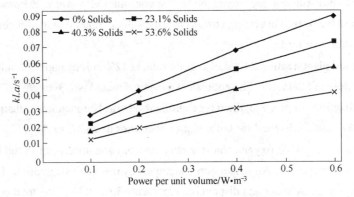

Fig. 5.2 Effect of solids concentration on oxygen mass transfer

The optimum operating slurry density must be determined from pilot test work the process results providing the required agitator specifications for the commercial process.

The relationship between the rate of oxygen demand and the oxygen mass transfer coefficient ($kl.a$) may be expressed as:

$$R = kl.a(C^* - C_L) \tag{5.12}$$

where R is rate of oxygen demand, mg/(L·s); C^* is saturated dissolved oxygen concentration, mg/L; C_L is dissolved oxygen concentration in the bulk fluid, mg/L; $kl.a$ is oxygen mass transfer coefficient, L/s.

The oxygen mass transfer coefficient for air dispersion in a mechanically stirred tank can be correlated with agitator power and superficial gas velocity by the following empirical correlation.

$$kl.a = c(P_g/V)a(u_s)^b \tag{5.13}$$

where P_g is aerated impeller power draw, W; V is unaerated fluid volume, m³; u_s is superficial gas velocity, m/s; c, a, b are empirical constants.

Typically the constant c is in the range 0.01-0.04 and the constants a and b are in the range of 0.3-0.6.

5.4 Operations Conditions and Process Requirements

5.4.1 Bioreactor configuration

The optimum configuration of reactors for a biooxidation plant is related to the rate of sulfide mineral oxidation and the corresponding rate of bacterial growth achieved. Sufficient residence time must be allowed in the primary bioreactors to allow a stable bacterial population to develop. If the residence time is too short, bacterial washout will occur and the rate of sulfide oxidation will decrease. In the case of biooxidation of the Sao Bento concentrate, a high specific rate of sulfide oxidation was achieved (15kg/(m^3·d)). By modeling the results of continuous pilot testwork, it was found that optimum process performance may be achieved by operation of four bioreactors in series to give an overall design retention time of 1.8d, or 0.45d per reactor, for a plant treating 150t/d of concentrate with the primary tank retention being 25% of the total plant retention.

The effect of tank configuration on process performance is illustrated for biooxidation of the Fairview concentrate. A low primary tank volume will result in a significantly lower sulfide oxidation for a given plant retention time, thus limiting the feed tonnage to the plant.

The optimum reactor configuration will be dependent on the ore mineralogy, and the rate of sulfide mineral oxidation and bacterial growth rate. Ores containing minerals which oxidize rapidly will allow biooxidation operation with a short retention time. The influence of the ore mineralogy on the rate of biooxidation is further discussed below. Fig. 5.3 is a kind of rotating drum bioreactor. It is found that the rotating drum bioreactor shows excellent capacity for the treatment of slurry with high solids concentration.

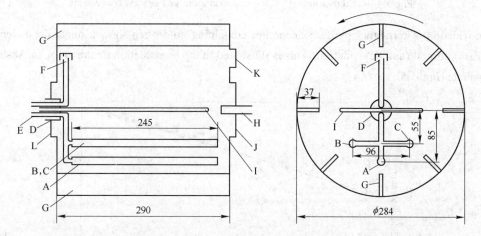

Fig. 5.3 Schematic diagram of the rotating drum bioreactor and main dimensions (mm)
A, B, C—gas-sparger; D—stationary shaft; E—gas inlet; F—gas outlet; G—baffle; H—probe (pH or DO);
I—heat exchanger; J, K, L—opening

5.4.2 Rate of sulfide mineral oxidation and gold dissolution

Pyrrhotite is readily oxidized during biooxidation by ferric leaching, according to the reaction given by Eq. 5.4. Sulfur produced will be partially oxidized by bacterial oxidation to produce sulfuric acid. Treatment of the Sao Bento and Fairview concentrates has shown that arsenopyrite is oxidized prior to pyrite and at a faster rate. There is the difference in rate of leaching for the sulfide minerals pyrrhotite, arsenopyrite and pyrite in the Sao Bento concentrate.

In sulfide mineral concentrates such as the Fairview concentrate, biooxidation will achieve complete release of locked gold with only partial oxidation of the sulfide minerals present. This effect is caused by preferential oxidation of arsenopyrite in which the bulk of the gold is "chemically bound". Research has also demonstrated that bacterial oxidation will occur preferentially along grain boundaries and in arsenic-rich zones, resulting in liberation of gold which is concentrated in these areas. This effect is shown in the biooxidation of Fairview concentrate (Fig. 5.4), in which preferential oxidation of arsenopyrite results in the early liberation of locked gold.

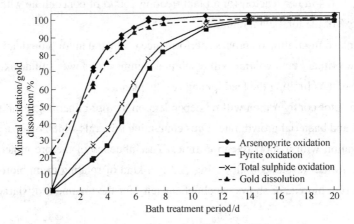

Fig. 5.4 Gold dissolution vs. mineral oxidation for Fairview concentrate

Biooxidation of refractory pyrite concentrates containing gold often show a linear dependence of gold dissolution on sulfide oxidation. This is illustrated in the biooxidation treatment of an Australian concentrate (Fig. 5.5).

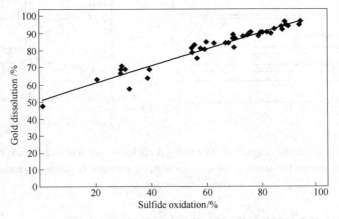

Fig. 5.5 Gold dissolution vs. sulfide oxidation for an Australian concentrate

5.4.3 General process requirements

5.4.3.1 Grind size

In general, the optimum grind size of 80% passing 75mm and 95% passing 150μm is stipulated for BIOX® plant operation. An increase in grind size reduces the sulfide oxidation rate and would result in a lower overall sulfide oxidation in a fixed volume plant. Production of a large proportion of fines (minus 25μm) may enhance the sulfide oxidation rate. However, down-stream thickening and filtration may become increasingly problematic and viscosity of CIL slurries will increase.

5.4.3.2 *Nutrients*

The nutrients potassium, nitrogen and phosphate are required for bacterial cell growth. Nutrient addition used originally in 1987, for the Fairview plant, was as follows; in kg/t of concentrate: 9.62 NH_4^+, 13.05 K^+ and 1.5 PO_4^{3+}. This was reduced over a period of two years to 8.40 NH_4^+, 1.42 K^+ and 1.56 PO_4^{3+}.

The minimum amount of nutrients required will be specific to the ore being treated. Often potassium occurs in silicate minerals present in the ore which are leached during biooxidation so that the required addition of potassium may be eliminated or reduced to a level of less than 1kg per tonne of concentrate.

5.4.3.3 *Toxins*

Biooxidation relies on a living culture of bacteria to promote sulfide mineral oxidation. Certain chemical reagents and naturally occurring elements are toxic to the bacteria and must be prevented from entering or building up to toxic concentrations, in the biooxidation plant.[4] Reagents used upstream of the biooxidation plant must be cleared by toxicity testing at the relevant dosages. These reagents include flotation reagents, flocculants and defoamers.

Reagents which are particularly toxic include:

(1) All cyanide and thiocyanate species above specified operating levels;

(2) Bactericides and fungicides;

(3) Descalants and corrosion inhibitors.

Maximum thickening of the flotation concentrate prior to use as BIOX® feed is very important in order to minimize ingress of flotation reagents into BIOX®, and prevent the concomitant froth formation and possible inhibitory effects on bacterial oxidation of sulfide minerals.

Toxic or inhibitive effects of soluble inorganic species are discussed in the sections on the BIOX® bacterial culture (earlier) and the effects of chloride and arsenic (to follow).

5.4.3.4 *Material selection*

It is normal practice to construct the BIOX® reactors and agitator impellers from stainless steel (304L, 316L or SAF_{2205}), or from mild steel with rubber lining, protection against corrosion and wear. Stainless steel is used for construction of cooling coils, with series 304L or 316L normally being adequate grades for corrosion resistance under BIOX® operating conditions. However, higher grades such as SAF_{2205} or 904L are required where dilution water contains high chloride concentrations (above 1,000 ppm Cl) as experienced in certain areas of Western Australia.

5.4.4 Effects of chloride and arsenic on BIOX® process

High concentrations of chloride in solution have been shown to cause damage to the membranes of the bacteria. Toxicity tests in which the rate of ferrous iron oxidation by the BIOX® bacterial culture was determined at different solution concentrations of chloride, have shown that little inhibition occurs at concentrations in the range 0-5g/L chloride. At concentrations of 7g/L chloride, the rate of ferrous iron oxidation decreases, such that only 55% oxidation occurs after 24h, compared to 80% for the control tests (0.1g/L chloride) and complete oxidation is achieved within two days. Above 19g/L chloride, complete inhibition of bacterial activity occurs (GENCOR, unpublished).

The bacteria in the BIOX® bacterial culture may operate at chloride concentrates in solution of up to 5g/L and maintain adequate sulfide mineral oxidation rates. However, the presence of high chloride concentrations in solution during biooxidation will promote precipitation of jarosite. This may become excessive and inhibit gold recovery on subsequent cyanidation of the washed biooxidation product, due to the coating of liberated gold. The precipitation of jarosite may also inhibit sulfide mineral oxidation by coating the sulfide mineral surfaces, thus preventing direct bacterial attachment. Continuous biooxidation of a refractory gold ore concentrate may only give satisfactory results at a relatively low chloride concentration in solution. Experience with continuous biooxidation of an Australian concentrate showed that best results in terms of sulfide oxidation and gold recovery were achieved at a chloride level of 1.3-1.5g/L.

Operation at high chloride concentrations will also require the use of more expensive stainless steel for internal cooling coils and the protection of pipes and amp internals.

The BIOX® bacterial culture is tolerant to arsenic (V) concentrations of 15-20g/L for continuous biooxidation of sulfide mineral concentrates. However, the culture is less tolerant to arsenic (III) concentrations, as discussed earlier. This difference may be explained by the toxic effect of arsenic (III), which is believed to inactivate enzymes which play a key role in cell metabolism. Arsenic (V) is less toxic than arsenic (III) but its action interferes with oxidative phosphorylation, leading to energy loss in the bacteria.

Biooxidation of the Fairview concentrate results in an arsenic (V) concentration in solution of 12g/L. However, even in the primary bioreactors, the arsenic (III) concentration remains below 500ppm during continuous operation. It is believed that arsenic (III) is rapidly oxidized to arsenic (V) by ferric ions, at pyrite surfaces. Fairview concentrate has a low pyrrhotite content and consequently biooxidation operates at a high redox potential (550-600mV vs. Standard Calomel Electrode), which may promote oxidation of arsenic (III) to arsenic (V).

In the case of biooxidation of the Sao Bento concentrate, high concentrations of 3-6g/L arsenic (III) occur in solution. It appears that the pyrite content of the concentrate (10.1%) does not promote the oxidation of arsenic (III) to arsenic (V). This is believed to be due to the high pyrrhotite content and/or low pyrite arsenopyrite mass ratios in the Sao Bento concentrate. The mechanism of arsenic (III) formation and oxidation is currently being researched, to verify the interaction of solution potential and available catalytic surfaces, such as pyrite, on the rate of oxidation.

The presence of arsenic (III) in the biooxidation product liquors may also adversely affect the subsequent neutralization process, particularly if the arsenic (III) concentration is greater than 3g/L and the iron/arsenic mole ratio is less than 6/1. Oxidation of neutralization liquors by a strong oxidant such a hydrogen per oxide may be require, in order to convert arsenic (III) to arsenic (V) and produce stable solid residues.

5.5 Summary

BIOX® process has been established as a commercially viable process for the treatment of refractory gold ores, offering lower capital and operating costs compared to alternative process routes such as roasting and pressure oxidation.

The neutralized effluent produced from the BIOX® process may be safely disposed of to a tailings dare and conforms to the U.S. Environmental Protection Agency specifications regarding solubility of arsenic. The demand for BIOX® technology continues to grow and several new commercial plants are expected to be constructed in the near future.

Questions

1. What is the character of the BIOX® process?
2. What is kind of bacterium used in the BIOX® process?
3. What are the main chemical reactions in the BIOX® process?
4. How the variables affect the oxidation and chemical leaching of principal refractory gold ore minerals?
5. Why the optimum grind size is stipulated for BIOX® plant operation?

Note
注　释

❶ 由于各种因素影响细菌群的组成，如高温、低 pH 值会使螺旋菌数量增加，因此使用混合菌时一定要确定菌群的性质。

❷ 溶液中较高浓度的亚铁离子可以促进反应器中液相里细菌群的增加，减少菌群失效的可能性，但较高浓度的亚铁离子也会降低氧化还原电位从而改变氧化反应过程。

❸ 硫化物的氧化是一个放热反应，而嗜温菌氧化反应的合适温度为 30~45°C，所以为了硫化物矿物有较高的氧化率，必须对生物反应器进行降温以保证适当的温度范围。

❹ 生物氧化过程的效率依赖于细菌的存活状态，某些化学试剂或天然元素对细菌来说是有害的，所以在氧化时一样要避免其他有害杂质的进入。

6 Biohydrometallurgical Processing of Colt, Nickel and Zinc

本章要点 本章简要介绍提取金属钴、镍和锌的生物冶金工艺方法。采用氧化亚铁硫杆菌浸出辉钴矿中的金属钴，超过95%的钴可被成功提取。当添加适应Ni的菌种后，氧化过程变为直接进行，但发现部分氧化亚铁硫杆菌被NiS吸附。可先由嗜酸菌氧化硫化锌，然后采用溶剂萃取或电沉积从含锌溶液中提取金属锌。

6.1 Introduction

As stricter environmental laws dealing with control of atmospheric, water, and soil pollution that results from conventional extraction by pyro- and hydr-ometallurgy are enacted, many processes of base metal extraction by bioleaching currently viewed as unattractive economically are likely to become competitive with the conventional processes, especially if the efficiency of these bioleaching processes can be further improved.

In this chapter, the advances in bioleaching of a series of base metals other than copper will be reviewed. These processes include some that are based on metal extraction by autotrophic bacteria, in particular iron-oxidizing acidophiles, and others that are based on extraction by heterotrophs, including aerobic and anaerobic bacteria, and fungi, which are generally aerobes. Leaching by acidophilic autotrophs has the advantage of requiring supplementation of the culture medium with only small amounts (concentrations in a gram per liter or less) of one or a few inexpensive, inorganic nutrients, especially in reactor leaching. In in situ-, heap- or dump- leaching of sulfidic ores by acidophiles, nutrient supplementation is mostly unnecessary because all needed nutrients are present in the environment in sufficient quantities.

Heterotrophic leaching, on the other hand, requires the addition of significant quantities (concentrations in the range of 10-100g/L or more) of an organic energy/carbon-source and an inorganic and/or organic nitrogen source in the lixiviant to support the growth and activity of the leaching microorganisms. The energy/carbon source may be carbohydrate such as sugars or other polysaccharides, proteins, alcohols, organic acids, aliphatic or aromatic hydrocarbons, heterocyclic compounds, or others. The choice is governed by the ability of an organism to metabolize a given energy/carbon source, its ready availability and cost, and its potential for exerting a selective effect for the organism(s) that are the active agents in a leaching process. Unlike leaching processes with acidophilic autotrophs, in which the very acidic growth conditions (pH range from 1.5-2.5) generated by the organisms exert a highly selective effect, leaching with heterotrophs may require different controls to suppress growth of undesirable organisms. The undesirables may consume the energy/carbon source without extraction of base metal from an ore mineral. One kind of selective control may involve the use of a specialized energy/carbon source that is unavailable or toxic to undesired organisms but readily used by the leaching organism(s), e.g., phenol. Another control would be to limit the access of oxygen if the leaching organisms were microaerophilic or anaerobic. Temperature and pH controls are two other alternatives for maintaining selective growth conditions. Elimination of interference by undesirable organisms may also be achieved with two reactors running in tandem. This approach is applicable when direct contact between the leaching organism and the ore to be leached is not required because the leaching organism produces one or more water-soluble compounds when metabolizing the energy/carbon source and these compounds do the leaching. In a two-reactor system, the organism generates the lixiviant axenically in one reactor, and the spent growth medium is then allowed to leach the ore in a second reactor or in a heap or dump without special precautions against contamination. The lixiviant can thus be generated without interference from contaminants. Since the spent medium containing the lixiviant may not support growth of interfering organisms, aseptic conditions become unnecessary for running the second reactor.

6.2 Cobalt Bioleaching with Autotrophic and Mixotrophic Bacteria

Besides copper and iron, other base metals, among them Co, Ni, Zn, Pb, Mo, and Ga, may occur in industrially significant quantities in sulfidic ores. Silver and platinum-group metals (platinum, rhodium, ruthenium, palladium, osmium, and iridium) may also be associated with sulfidic ores. In some instances, these metals may be the chief ore constituent, whereas in other instances they occur in polymetallic ores. Laboratory studies have shown that sulfidic ores containing these metals can be leached by acidophilic iron bacteria such as *Thiobacillus ferrooxidans*, either by an indirect or a direct mechanism, or both. In an indirect mechanism, the organisms generate an acidic ferric lixiviant which oxidizes the metal sulfide(s), whereas in a direct mechanism, the organisms oxidize the metal sulfide in the ore directly while in contact with the mineral surface and interacting enzymatically with its crystal lattice. In extracting polymetallic ores, galvanic effects can sometimes be exploited for the preferential leaching of a desired metal constituent.

The chemistry underlying the leaching of the base metals is the oxidation of the sulfide moiety to sulfate. In chemical oxidation of sulfide by ferric iron, the sulfide is converted to elemental sulfur (S^0). This sulfur may accumulate on the mineral surface and prevent further oxidation by blocking further access of ferric iron to the sulfide mineral. However, organisms like *T. ferrooxidans* and *T. thiooxidans* readily oxidize so to sulfate and thereby sustain chemical metal sulfide oxidation. With the exception of $PbSO_4$ and jarosite, a basic ferric sulfate, all base metal sulfates are soluble in the acid lixiviant and thus do not precipitate on the surface of the sulfide mineral from which they derive. The bioleaching of galena (PbS), requires special conditions to prevent sparingly water-soluble $PbSO_4$ build-up on the galena surface. In the laboratory, continual agitation on a shaker has been found to give satisfactory results.

Several studies on Co extraction by acidophiles were chiefly concerned with testing the amenability of different cobalt-containing raw materials to leaching. Thus, Torma et al investigated bioextraction of cobalt from cobaltite (CoAsS) oncentrate from the Blackbird Mine in Idaho (USA) by *T. ferrooxidans*. The chief mineral constituents of the ore were pyrite, arsenopyrite, cobaltite and silica. The ore concentrate contained 44.3%total S, 32.7% Fe, 9.9% As, 5.7% Co, and 1.0% Si. Leaching was performed in batch experiments in 250mL Erlenmeyer flasks containing iron-free mineral salts solution plus the ore concentrate at a desired pulp density (PD). Experimental flasks were inoculated with iron-grown *T. ferrooxidans*. Ore concentrate at PD of 3%-5% gave best results. In 42 days, 98% of the Co was extracted at 3% PD, and 78% at 5% PD. At higher pulp densities, total Co extraction in the same length of time was markedly smaller; e.g., at l0% PD, it amounted to only 13.2%, and at 30% PD to just 4.5%. Leached Co was recoverable from the pregnant solution by solvent extraction with 15% (v/v) cyanex 272 or cyanex 302, and 85% (v/v) Exxsol D-80 after precipitation of Fe and as basic ferric jarosite and ferric arsenate, respectively. Co was stripped from the extractants with dilute sulfuric acid.

In a comparison of 29 strains of *T. ferrooxidans* in leaching of Blackbird cobaltite ore and concentrate, and synthetic cobalt sulfide, Thompson et al. found 5 strains to give the best extraction. Optimal leaching without added iron occurred at an initial pH of 2.0. However, in lixiviant containing ferrous iron, optimal leaching occurred at an initial pH of 1.8. More than 95% of the cobalt in ore concentrate was extracted in 28 days by wild-type strain Fe in the initial presence of Fe^{2+} of 1500 mg at a PD 1%. Leaching was found to occur by direct and by indirect attack (with Fe^{3+}) of the ore or its concentrate. Most of the nitrogen and phosphorus requirements of the bacteria were met by the ore. When oxidation of synthetic CoS by strain T5 was studied, 80% of the cobalt became dissolved in 200h while only 8% was dissolved in sterile controls. The mechanism of action in this case appeared to be mostly by the direct mode.

Morin et al. performed bench-scale leaching experiments on a cobaltiferous pyrite concentrate with a mixed culture of *T. ferrooxidans*, *T. thiooxidans* and *Leptospirillum ferrooxidans*. The cobalt was finely disseminated in the pyrite matrix and amounted to 1.4% of the total mineral. Leaching was performed in a modified 9K medium. In batch tests in airlift reactors, optimal leaching occurred when the pH was maintained in a range of 1.1-2.0. A build-up of ferric iron in excess of 35g/L was found to be inhibitory to pyrite oxidation. Addition of CO_2 at 1% of the air supply (source of oxygen) stimulated bacterial activity. Optimal Co extraction was observed at a particle size of < 20µm and a pulp density of 10%. Up to 80% of the pyrite was oxidized. When the same ore was leached in a reactor system consisting of a set of four 20L leach tanks connected in series that were continuously fed with a medium similar to that in the batch experiments, and in which the spent medium was recycled, a cobalt concentration of 5 g/L in the pregnant solution was achieved, with 80% of the Co extracted from the ore. Prolonged operation of the reactor system with a gradual increase in pulp density of the ore from 10%-20% permitted sustained Co extraction. The pH during reactor operation was maintained between 1.3 and 1.7 with addition of limestone pulp. Not surprisingly in view of the low pH, the authors reported that *L. ferrooxidans* displaced *T. ferrooxidans* and *T. thiooxidans* in the second and third tanks of a four-tank series. *L. ferrooxidans* replaced the other two organisms essentially completely in the fourth tank, in which the pyrite oxidation rate was very slow.

Baldi et al. reported that small amounts of calcitic gangue (0.01%-1.01%) in pyrite ore can interfere in its oxidation by *T. ferrooxidans* and in the accompanying release of cobalt and zinc that may be present in the ore. The reason for the interference is the neutralization by the calcite of acid required for growth of *T. ferrooxidans*. Continual addition of small amounts of acid (5mmol/L H_2SO_4) to ore in percolator setups removed the calcite. *T. ferrooxidans* began to grow and oxidize the ore with release of Fe, Co and Zn once the calcite was removed. The lag phase depended on calcite content of the ore and the time required for its removal. Although leaching rates were greater in percolator setups than in shake flasks, the duration of the lag phase was shorter in shake flasks than in percolators.

Nakazawa and Sato took the unusual approach of using *T. ferrooxidans* in leaching Co from cobalt-rich ferromanganese crusts, which have no sulfidic constituents. The leaching of the crusts was accomplished in 9K medium with elemental sulfur or pyrite (−100mesh) replacing ferrous sulfate as the energy source or the bacteria. The crust contained 20.8% Mn, 13.8% Fe, 0.82% Co, 0.5% Ni, and 0.12% Cu. Leaching was performed at 25°C in shaken flasks with crustal material (−100mesh) at a pulp density of 0.1%. Although acidification with H_2SO_4 to pH 1 in the absence of bacteria released nearly all Cu and some of the Ni and Fe in 18 days, no Co or Mn were released under these conditions. With the addition of 1% elemental S to Fe-free 9K medium, the bacteria leached all Cu, nearly all Ni, Co rid Fe, and more than 60% of the Mn in 18 days. Raising the sulfur concentration in the medium to 3% speeded up Co leaching significantly, but the full extent of leaching remained nearly the same (between 80% and 90%) as with 1% sulfur. Use of a copper-tolerant strain of *T. ferrooxidans* improved the kinetics of Co and Ni leaching.

Olson et al reported successful leaching of cobalt from smelter wastes (sulfidic dross furnace matte containing Co and Ni) by *T. ferrooxidans* in batch experiments. In six weeks, they were able to solubilize two-thirds of the 8.5% cobalt in the matte at a pulp density of 4% and a mesh size of −200 (<74μm) in the basal salts solution of 9K medium. At a mesh size of −50 +100mesh (150-300μm), only 43% of the Co was leached. Unlike Co, the Ni in the matte was leached by acid alone. Addition of pyrite did not increase the rate or extent of Co leaching.

6.3 Nickel Bioleaching with Autotrophic and Mixotrophic Bacteria

Some studies on nickel extraction from nickel sulfide and sulfidic nickel ores and ore tailings by acidophiles have appeared since 1990. One such study by Amaratunga et al explored the bioextraction potential of ore tailings for heap and vat leaching. Pyrrhotitic, nickel-containing (0.7%-0.8%) tailings material from Falconbridge, Ontario, Canada was subjected to leaching by a strain of *T. ferrooxidans* in 9K salts medium in shake flasks at a pulp density of 8%-10%. As much as 50% of the Ni was extracted in 28 days in the presence of *T. ferrooxidans*. It was suggested that agglomerating the tailings with gypsum hemihydrate (flue gas desulfurization gypsum) prior to leaching might minimize segregation of fine clays in the tailings, which could cause channeling in leach heaps and reduce solution contact with ore particles, but no test results were presented.

Kai et al. studied the effect of enhancement of nickel tolerance to 1g/L in a strain of *T. ferrooxidans* on nickel solubilization from NiS. The original strain was isolated from acid mine water of the Yahara Mine in Japan. Tests were run in batch experiments in 500mL Erlenmeyer flasks containing 300mL of a ferrous iron (8g/L) solution at pH 2.0 and 0.1g of reagent-grade NiS and shaken at 30°C. About 40%

nickel was extracted by the adapted strain in 150h compared to 30% by the unadapted strain; however, about 25% of the Ni in the NiS was solubilized in a sterile control. No difference in the rate of Fe^{2+} oxidation was noted between the Ni-adapted and unadapted strains. From this the authors inferred that NiS oxidation by the unadapted strain was indirect (due to oxidation by Fe^{3+} produced by the bacteria) whereas the additional NiS oxidized by the adapted strain was the result of direct attack. The authors reported that 30%-50% of the cells of *T. ferrooxidans* were adsorbed to the NiS.❷

Behera et al. study the bioleaching of nickel and cobalt from lateritic chromite overburdens used by organic acid of *A. niger*. TEM analyses were carried out with the samples collected before and after leaching. It was shows that nickel dissolute into the solution leaving a porous space in the matrix of the hematite by forming nickel oxalate or nickel citrate, in Fig. 6.1.

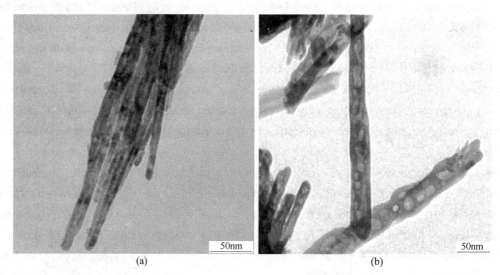

(a) (b)

Fig. 6.1 TEM analysis of pre-treated chromite overburdens

(a) Before leaching; (b) After leaching

6.4 Zinc Bioleaching with Autotrophic and Mixotrophic Bacteria

Zinc sulfide (sphalerite) has been shown by various investigators to be oxidizable by iron-oxidizing acidophiles like *T. ferrooxidans*. Among studies since 1989, one by Krafft and Hallberg assessed the possibility of in situ mining of zinc sulfide ores from two different Swedish mines, Saxberget and Kristineberg. Ore from Saxberget mine contained 15% Zn and 1.5% Cu, whereas that from the Kristineberg mine contained 8% Zn and 2.5% Cu. Leaching studies were performed in columns and flasks using mixed cultures from the Falu copper mine in Sweden and the Rio Tinto mine in Spain. The dominant species in these cultures was *T. ferrooxidans*. The culture media were various modifications of 9K medium. In column leaching, only 0.84%-1.02% of the Zn and 1.8%-3.2% of the Cu in the Saxberget ore were led in 87 days, and only 0.2%-1.2% of the Zn and 0-1.2% of the Cu in Kristineberg ore. Highest leaching rates with respect to Zn were obtained with particles in the 16-32mm size-range. In batch-leaching experiments 80% of the Zn in either ore was leached in 150 days.

In an attempt to enhance the kinetics of selectively leaching the Zn from copper-zinc sulfide concentrate, Carranza et al. employed a two-phase system in which they generated ferric lixiviant with *T. ferrooxidans* in a first phase and then leached the ore concentrate with the lixiviant in one or more reactors in series in the second phase. The lixiviant-generating phase was operated at 31°C and the leaching phase at 70°C. Galvanic effects due to the presence of chalcopyrite with the zinc sulfide caused preferential leaching of Zn. Maximum Zn extraction of 80% at a pulp density of 12% was obtained in a continuous process. The authors predicted hat maximum Zn extraction can be achieved with mean residence times ranging from 8-22 h, depending on the number of reactors in series employed for the second phase. The zinc in the pregnant solution from this process was recoverable by solvent extraction and electrowinning.❸

Shi et al. investigated the bioleaching processes of zinc sulphides with *Acidithiobacillus ferrooxidans* (A.f.) and a moderately thermoacidophilic iron-oxidizing bacterium (MLY) by leaching experiments. Fig. 6.2 shows the bioleaching process had the more rapid reaction kinetics in the presence of MLY, compared with that by the bacteria A. f.

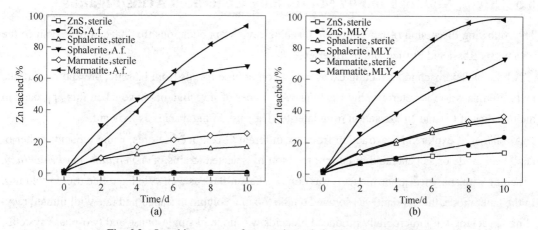

Fig. 6.2 Leaching curves of marmatite, sphalerite and synthetic ZnS

(a) *Acidithiobacillus ferrooxidans*; (b) a moderately thermoacidiophilic microorganism (MLY)

Different leaching conditions can affect the rates and extents of Zn extraction from Zn ores. Ballester et al. compared three strains of *T. ferrooxidans* adapted to $FeSO_4$, $CuFeS_2$ and ZnS, respectively, for their ability to oxidize ZnS in 9K medium. The two strains adapted to the sulfide ores solubilized more Zn from an ore concentrate and without an initial lag than did the $FeSO_4$-adapted strain. When three zinc-ore concentrates differing in the amounts and kinds of iron impurities were leached with the ZnS-adapted strain of *T. ferrooxidans*, best results were obtained with a concentrate in which ore impurities were dissolved in the sphalerite (ZnS). Next best results were obtained with a concentrate comprised mainly of sphalerite and pyrite out nevertheless significant amounts of Zn, were dissolved from a concentrate comprising sphalerite, pyrite and galena. These differences may be at least in part explainable in terms of galvanic effects. Ballester et al. also suggested that Fe^{2+} when dissolved in ZnS enhances the conductivity of ZnS and its chemical leachability. In comparing leaching rates at 5%, 10% and 15% PD of ore concentrate, fastest leaching occurred at the 5% PD and slowest at 15% PD. Ballester et al. suggested that the lower pulp density facilitated O_2 and CO_2 transfer in the reactor system. They also found that progressively more Zn was dissolved as the reactor temperature was increased from 25-30°C and then to 35°C, although initial rates at 30°C and 35°C were similar. In bioleaching, the sphalerite was preferentially attacked at the site of cracks and structural defects.

Sukapun et al. were able to extract Zn from zinc silicate ore from Thailand using acid ferric sulfate lixiviant generated by *T. ferrooxidans*. The action was demonstrated in shake flasks and columns. In shake flasks at a pulp density of 10%, about 60% of the Zn in the ore was extracted in 20 days, during which the bacteria generated ferric iron from ferrous iron. Ferrous iron leached as much as 40% of the zinc in the ore, presumably by displacing it. At a pulp density of 2%-5%, the bacteria mobilized almost 75% of the Zn in the ore. In column experiments less than 20% of the Zn in the ore was mobilized. In these experiments the medium containing the culture was continually recycled. Control of pH was critical to maintaining bacterial activity in the columns.

6.5 Metal Mobilization by Microbially Generated Acids/Ligands

The following discussion of heterotrophic leaching reviews publications that have appeared since the reviews by Rossi and Ehrlich and by Brierley.

A possible approach to extracting Co as well as other base metals from lateritic, ore and nonsulfidic metalliferous waste is heterotrophic leaching. A number of investigators have taken this approach in their studies of Co and Ni extraction from laterite. They used fungal acids as lixiviant.

Alibhai et al explored Ni extraction from six different Greek lateritic ores, using bioacids either of analytical grade or as generated in culture medium by selected strains of *Aspergillus* and *Penicillium*. Leaching with analytical grade bioacids was done in shaken-flask as well as in fixed-bed trickle columns. In the latter mode, the lixiviant was sprayed on the top of a column at a specified rate with limited recycling. Leaching with microbially produced bioacids was done in single-phase and two-phase systems. In the two-phase mode, the bioacids were generated in one reactor, and the resultant spent medium was then applied to ore in a second reactor. In the single-phase mode, the bioacids were produced by the fungi in the presence of the ore. In leaching with purified bioacids in the shaken mode, up to 40% of the Ni in low-grade ore was extracted in 12 days, and 55% of the Ni in high-grade ore in 9-15 days. In the column mode, >30% of the Ni in low-grade ore was extracted with citric acid after 100 days. Sulfuric acid was somewhat more effective than citric acid but both acids were more selective for Ni than for Fe. Oxalic acid extracted Fe in preference to Ni. Its presence is thus not desirable among fungally produced bioacids used as lixiviants for nickeliferous laterites. Leaching with bioacids generated in flask cultures of *Aspergillus* and *Penicillium* strains was performed in one-phase and two-phase systems. The authors reported that 68% of the Ni in low-grade Litharakia ore (0.73% Ni; 0.043% Co) was leached in 51 days in two-phase system, using an *Aspergillus* Strain in a sucrose-containing medium. This result was attributed to optimal citrate production. Tests in single-phase mode were expected to give somewhat better results than tests in two-phase mode, but were not reported in detail by these authors. Detailed results from such tests with a different ore were reported by Kar et al.

A more detailed study of the action of analytical-grade organic acids was undertaken by Tzeferis and Agatzini-Leonardou. The action of these organic acids on high-grade Kastoria laterite (4.23% Ni; 0.028% Co) and low-grade Litharakia ore (0.73% Ni; 0.043% Co) singly and in combination with and without added H_2SO_4 was compared to that of H_2SO_4 alone. The organic acids tested included citric, lactic, formic, oxalic and salicylic. Leaching tests at temperatures up to 50°C were carried out in stirred conical flasks in a water bath, whereas tests from 50-95°C were carried out in spherical, stirred glass reactors heated by a thermostatically controlled thermal mantle. Lixiviant concentrations, pulp densities, temperatures, and reaction durations were varied for each of the different lixiviants. Of all the organic acids tested, citric was the most effective (max. extraction ~60%), but equimolar concentrations of citric and sulfuric acid were in some instances slightly note effective. Reaction kinetics for citric acid were slower than for sulfuric acid. Citric acid lixiviant whose acidity was adjusted with sulfuric acid to a free acid concentration of 0.5g/L was more effective than an equivalent concentration of citric acid whose free acid concentration was not adjusted, or sulfuric acid alone. This effect was explained in terms of the chelation of Ni by citric acid. Oxalic acid was found to extract iron preferentially. It liberated very little Ni in the tests. The extent of maximum Ni extraction from different ores under identical conditions varied.

Experimental evidence for the greater effectiveness of the single-phase mode versus the two-phase mode in extracting Ni from lateritic ore was presented by Kar et al. They found that more cobalt was extracted from lateritic nickel ore from India when the fungus *Rhizopus arrhizus* produced lixiviants in the presence of the ore than when using lixiviant consisting of culture medium in which the fungus had grown in the absence of the ore and from which the fungus had been removed by filtration after its growth. They attributed this difference in effect to higher local concentration of the bioacids formed in the presence of the fungus than in fungus-free spent medium. Their culture medium was potato-dextrose broth. The experiments were performed in 250mL conical flasks on a shaker at 37°C and 120 rev/min. After 20 days, 73% of the cobalt and 4.5% of the Ni were extracted from -350 mesh laterite containing 0.039% Co and 1.1% Ni at a pulp density of 5%. The acids produced by R. arrhizus in this study were not identified but can be assumed to have functioned as protonation agents in the displacement of cobalt and nickel from the ore and/or as ligands of these metals. On extended incubation, the fungal mycelium probably bound some of the solubilized metal.

Alibhai et al. reported on Ni extraction from high-grade laterite ore (Kastoria deposit, Greece) separately with acetic, formic, lactic, oxalic, and citric acids. Each of these can be microbiologically formed as end products of energy metabolism. Of these acids, 1.5 mol/L citric acid gave the best results with 40% Ni extraction in 20 days at 5% pulp density By contrast, 1.5 mol/L H_2SO_4 extracted 60% of the Ni under the same conditions. Some fungal strains of *Penicillium* and *Aspergillus* extracted 50%-60% of the Co and Ni from low-grade Litharakia ore, which contained 0.92% NiO, 0.060% CoO in sucrose-mineral salts medium. From 2.5%-3.5% of the Ni was adsorbed by the fungus mycelium. When molasses replaced sucrose in the medium, 20% less Ni was extracted from the ore.

Tzeferis et al. performed similar bioleaching experiments with the low-grade nickel ore from Litharakia. They also used strains of the *Penicillium* and *Aspergillus* in dextrose-mineral and sucrose-salts media at an incubation temperature of 30°C. The ore had a mesh size of −150 (<106μm). The ore sample used in this study had a nickel content of 0.73% but the Co content was not reported. Cultures were shaken at 400 rev/min. They found 60% of the Ni and 50% of the Co in the ore to be extracted in sucrose medium in 48 days, but much less in dextrose medium. A portion of the leached Ni became bound to the fungal mycelium. The lixiviant produced by the fungi contained citric and oxalic adds. Citric add production was much better with sucrose as carbon/energy source than with dextrose. Tarasova et al reported that direct ultrasonic treatment at a frequency of 18-20 kHz and an intensity of 1-9 W/cm^2 as a 30s or 120s pulse at the beginning of a leaching run with *Aspergillus* or *Penicillium sp.* improved Ni extraction from low-grade laterite ore from Greece from 40%-50%. The authors suggested that the ultrasonic energy uncovered occluded minerals in the ore and/or opened channels that facilitated access of the lixiviant to the valuable minerals. This treatment was, however, seen as too costly to apply commercially in the bioleaching of low-grade nickeliferous laterite.

Tzeferis studied the bioleaching of a low-grade laterite ore (−150 mesh or <106μm particle size) from LARCO S.A. (Larymna, Greece) in a two-phase system using a molasses-based medium. The ore contained 0.73% Ni and 0.043% Co. He generated lixiviant (bioacids, mainly citric and some oxalic) in a molasses-based medium with citric acid-producing *Penicillium* and *Aspergillus* strains in air-lift fermenters over eight days. The molasses was pretreated with potassium ferrocyanide to control the concentrations of Fe, Cu, Ca, and Mn, which affect citric acid production by the molds. The ore was then leached with the spent medium containing the bioacids in spherical stirred-glass reactors at 95°C. Maximum Co recovery from the ore ranged from 44%-49% and maximum Ni recovery from 54%-62%. Since molasses is cheaper than sucrose (cane or beet sugar) or dextrose, its use in leaching Co and Ni from laterite on an industrial scale is more economic. However, its pretreatment for control of Fe, Cu, Ca, and Mn levels in the fungal culture medium to maximize organic acid production is essential.

Sukla and Panchanadikar bioleached a nickeliferous laterite from Sukinda, India with an *Aspergillus niger* strain isolated from the ore. Leaching was carried out on 2g of the ore (particle size of −150 mesh) in 250mL Erlenmeyer flasks containing 100mL of a medium that included dextrose (2%-10%) and 20% potato extract (single-phase system) on a rotary shaker at 37°C. The ore contained 1.1% Ni, and 0.03% Co. Unlike experiments by some other investigators who found sucrose more effective than dextrose for the production of bioacids, Sukla et al found that with 10% dextrose, 92% of the nickel and 34% of the cobalt were extracted in 20 days at an initial pH of 6.5, which fell to 1.75-2.0 after 10 days.

Pradhan et al. reported that the leaching experiments were carried out to extract nickel, vanadium, and molybdenum present in spent refinery catalyst. A bioleaching process was applied in the first step. Pulp density, initial Fe (II) concentration, initial pH, particle size, and temperature were varied to optimize the bioleaching process.[❶] Ni, V, and Mo were leached out with maximum recoveries of 97%, 92% and 53%, respectively, at an optimized bioleaching condition of initial ferrous ion of 2 g/L, initial pH of 2, pulp density of 10% (w/v), particle size of 106-45 μm, and a temperature of 35°C. As the Mo leaching rate was very low, Applied with different concentrations of Na_2CO_3, $(NH_4)_2CO_3$, the leaching results of Ni, V, and Mo were shown in Fig. 6.3 and Fig. 6.4. The leaching efficiency of Mo increased with an increase in $(NH_4)_2CO_3$ concentration from 10g/L to 30g/L, and with no further increase. The leaching of Ni and V was similar to that achieved using Na_2CO_3. Hence, 47% of the Mo could be leached at a concentration of 30g/L $(NH_4)_2CO_3$, compared to 43% of Mo at a concentration of 40g/L Na_2CO_3. Therefore $(NH_4)_2CO_3$ was preferred as the leaching reagent over Na_2CO_3.

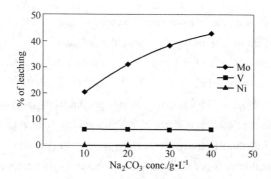

Fig. 6.3   Leaching of Ni, V, and Mo at different concentration of Na_2CO_3 for 1h

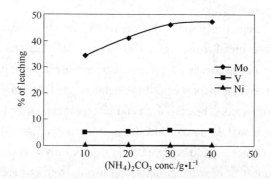

Fig. 6.4   Leaching of Ni, V, and Mo with different concentrations of $(NH_4)_2CO_3$ for 1h

Sukla et al. reported that in leaching Ni from nickeliferous laterite from Sukinda, India, with nickel-tolerant, autotrophic *T. ferrooxidans*, and heterotrophic *Bacillus circulans*, *Bacillus licheniformis* and *A. niger*, *T. ferrooxidans* was least effective and *B. circulans* most effective among the three bacterial cultures. However, *A. niger* was even more effective than *B. circulans*. Thus, in 20 days at a pulp density of 5% in appropriate growth medium for each of the organisms in shaken flasks, *T. ferrooxidans* leached 6% of the Ni in the ore, *B. licheniformis* 35%, *B. circulans* 85.4% and *A. niger* 92%. However, *A. niger* exhibited a lag of about 10 days before the rate of leaching became significant whereas *B. circulans* exhibited no such lag.

Zinc can be leached from some nonsulfidic ores and waste materials. Tungkaviveshkul et al. demonstrated Zn bioextraction from Zn silicate residue left after conventional processing of high-grade Zn silicate ore. They used *Aspergillus niger* ATCC 11414 and a strain of *Penicillium* isolated from the residue in their study. For Zn leaching, the cultures were grown in the presence of the ore at 10% pulp density in a sucrose-mineral salts medium. Both strains leached Zn from the ore. More than 60% Zn was extracted in 49 days, most of that in the first 14 days. *A. niger* produced citric, oxalic and tartaric acids, whereas *Penicillium* produced mainly citric and tartaric. The authors showed that this combination of acids was more effective in extracting the Zn than the individual acids. The fungal mycelium was found to adsorb some of the mobilized zinc. Some mobilized Zn was precipitated as Zn phosphate hydrazine in the culture medium. To maximize Zn recovery in solution, the culture conditions require further optimization.

The yeast *Brettanomyces lambicus* was shown to be able to leach 65% of Zn from filter dust residue of a copper refinery in a medium containing glucose and yeast-malt extract, 11.2% of the lead in a yeast extract-peptone medium, and 16.5% of the copper in a glucose-mineral salts medium. Twenty other yeast strains exhibited different abilities to leach Zn, Pb, and Cu from the filter dust. Best results were obtained at a pulp density of 0.5%. The leaching agents produced by the yeasts and implicated by the authors included lactic and acetic acids and possibly amino acids. The authors also reported that more metal was extracted by 17 of the 21 strains when they grew in the presence of filter dust than by their respective spent culture medium. It is noteworthy that *Brettanomyces lambicus* produced acid only in the presence of filter dust.

6.6 Summary

While the foregoing discussion clearly shows that microbes have a potential for assisting in extraction of base metals and some other metals from ores, either through their direct mobilization (bioleaching) or through removal of interfering ore constituents (biobeneficiation), economic exploitation of these activities on a commercial scale remains in most cases to be demonstrated. In many of the instances discussed here, the microbial activities need to be optimized and incorporated into industrial process designs before a meaningful economic analysis is possible. In most instances, rates of reaction at mesophilic or thermophilic temperature ranges need to be accelerated significantly over those observed in initial laboratory experiments, irrespective of whether autotrophs or heterotrophs are the biological agents. In the case of heterotrophs, an economically viable process requires cheap organic substrates and selective culture conditions that can be easily established and maintained on a large scale at minimum cost. Microbiological bioleaching or biobeneficiation processes of the ores discussed in this chapter will be competitive with conventional ore processing (pyro- or hydrometallurgy) if plant investments, costs of production, and control of environmental pollution are favorable, considering the extent and quality of the ore resource to be processed and the value of the product.

Questions

1. What is the difference between the autotrophic and mixotrophic bacteria?
2. What are usually the chief mineral constituents of the cobalt, nickel, and zinc ores?
3. How to enhance the kinetics of selectively leaching the Zn from copper-zinc sulfide concentrate?
4. What will affect the rates and extents of nickel extraction from nickel ores?
5. How does fungal acids as lixiviant work in the process of Co and Ni extraction from laterite?

Note

注 释

❶迟滞期取决于方解石矿物的成分和其被去除的时间,尽管在生物反应器中的浸出率远大于摇瓶,但迟滞期比摇瓶长。

❷采用适应 Ni 和不适应 Ni 的菌种得到的 Fe^{2+} 的氧化率没有明显差别,根据这个结果推断不适应 Ni 的菌种是间接氧化 NiS(是由细菌产生的 Fe^{3+} 进行氧化的),而添加适应 Ni 的菌种后,氧化过程变为直接进行。同时,实验中还测试发现30%~50% 的氧化亚铁硫杆菌被 NiS 吸附。

❸预测获得锌最大提取率的浸出时间平均为 8~22h,这取决于第二相串联反应器的数量。然后,采用溶剂萃取或电沉积可以将金属锌从含锌溶液中提取出来。

❹首先进行生物浸出,浸出过程中对矿浆浓度、初始 Fe(Ⅱ) 浓度、初始 pH 值、颗粒大小、温度等参数进行优化。

7 Biohydrometallurgical Recovery of Heavy Metals from Industrial Wastes

本章要点 冶金粉尘、矿物加工废弃物、河底淤泥、金属表面处理液等均含有重金属，如 Cu、Cd、Ni 和 Zn 等，其直接排放将危害环境和污染水源。本章简要介绍针对这些废弃物的特点，采用不同的细菌和生物冶金技术工艺，回收其中的重金属，以避免或减轻对环境生态的危害作用。

7.1 Introduction

In consequence of a high industrial expansion, large quantities of industrial wastes are accumulating in many countries and cannot be disposed without prior special treatments. In particular, waste products from the mining and metal refining industries, sewage sludges and residues from power station and waste incineration plants can contain heavy metals at high concentrations. For this reason, they cannot be disposed into wastewaters plant treatment and must be submitted to special treatment in order to reduce metals content. Besides, some waste products containing high concentration of metals (copper, zinc, lead, chromium, etc.), may be regarded as a secondary source of metals.

Heavy metals often occur as free elements in the environment. However, in many cases these metals are chemically encapsulated in minerals such as arsenopyrite, pyrite, pyrrhotite and chalcopyrite. This encapsulation represents a physical barrier that prevents the recovery of the metal ion via conventional methods such as cyanidation for gold recovery. Researchers in many countries have investigated the application of biotechnology in mining over the last 40 years. To date, several biotechnologies have been exploited commercially in well-mechanized, engineered systems that can be categorized under the term "biohydrometallurgy".

Biohydrometallurgy is the interaction between metals and microbes with the specific aim of converting insoluble metal sulfides to soluble metal sulfates.❶ Bioleaching has been defined as the dissolution of metals from their mineral sources by certain naturally occurring microorganisms or the use of microorganisms to transform elements so that the elements can be extracted from a material when water is filtered through it. For example, in the case of copper extraction, copper sulfide is microbially oxidized to copper sulfate and the metal ions are concentrated in the aqueous phase and the remaining solids are discarded. The closely related technology to bioleaching, known as "biooxidation" refers to the conversion of solid metals into their water-soluble forms using microbial processes. Therefore, "biooxidation" describes the microbiological oxidation of minerals, which contain metal compounds of interest. As a result, metal values remain in the solid residues in a more concentrated form. For example, in the recovery of gold from arsenopyrite ores, the gold remains in the mineral after biooxidation and is then extracted via cyanidation in a subsequent step.

The bioleaching process is largely documented in literature; on one hand it is applied to the extraction of metals from various ores and ore concentrates, on the other hand it is employed to treat wastewaters containing metals at high concentrations. The later aspect will be discussed in this chapter.

7.2 Metal Recovery from Waste Sludge and Fly Ash

The waste sludge under treatment was a dust coming from a metallurgical furnace for iron-manganese alloy production, and the sludge contained heavy metals of Cd, Cu, Zn, Pb, Ni, and so on. Solisio et al. reported that a biological solubilisation of the heavy metals was performed in batches employing a strain of *Thiobacillus ferrooxidans* (an acidophilic bacterium) that can either grow on reduced sulphur compounds or on ferrous iron. It utilises energy obtained from oxidation of inorganic sulphur compounds (e.g. Fe_2S, $CuFeS$) as well as ferrous iron dissolved in a liquid medium. Ferrous iron is oxidized to ferric iron in acidic medium by means of *T. ferrooxidans*, while its chemical oxidation by means of oxygen is extremely low.

Increased concentrations of the sludge (from 2% to 10%) were submitted to the bioleaching process in order to evaluate their effect on the metals solubilisation process: the removal of the selected metals from the sludge, as listed in Table 7.1. The increasing concentration of sludge allowed removal of about 73% of zinc, for each tested concentration (from 2% to 10% dry sludge). The trend of bioleaching is confirmed by data referring to pH and Fe(II) consumption. In fact when the percentage of metal removed is high (71%-76%), the pH remains low (pH 2-3) and Fe(II) is almost entirely oxidized to Fe(III).

Table 7.1 Results showing the effects of the sludge percentage after 9 days of bioleaching process

Sludge content/%	Fe^{2+}_{OX}/%	Zn removed/%	pH
1	96	76	2.0
2	94	73	2.5
4	95	73	2.5
6	87	71	2.5
8	85	72	3.1
10	99	73	3.0

The lag phase necessary to reach the maximum removed metal was completely dissimilar for sludge (Fig. 7.1). For the sludge tested, 2-4 days were sufficient to reach a significant percentage of removed metals. These behaviors can indicate the necessity of an adaptation period of the microbial strain to the different solid wastes. Further investigations are necessary in order to evaluate the factors that may hinder an efficient microbial growth.

The utilised wastes show a different affinity for the employed microbial strain; from the obtained results, the bioleaching process may be favourably employed for the sludge, because it solubilizes over 70% of zinc without adding sulphuric acid. Although the technical feasibility of the process has been potentially evaluate, it is clear that the economical advantages of the process must be verified considering all the economical factors present in the bioleaching process such as the availability of $FeSO_4$ supplied at low costs or as by-product.

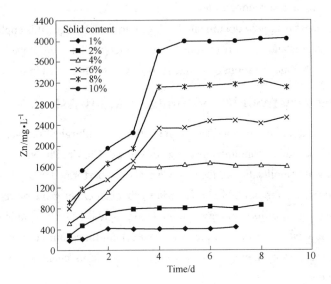

Fig. 7.1 Trends of zinc removal for increasing concentration of solid content

A bioleaching process using a savage microbial strain mainly containing *T. ferrooxidans* has been employed in the removal of heavy metals present in the waste sludge. At the initial waste concentration 1%, the percentage of removed metals was 76% and 78% for zinc and aluminium, respectively. Increasing the initial slurry concentration, the percentage of removed zinc was approximately 72%-73%, while a considerable decrease in the aluminium removal has been observed. In particular, at initial sludge concentration of 3%, only 15% of aluminium extraction was obtained in the investigated experimental conditions. Such different behaviour was confirmed also by an increasing of the lag phase that resulted 2-4 days for zinc solubilization and 12-14 days for aluminium removal.

In addition, two different kinetic models were studied for describing the metal dissolution in the investigated experimental conditions. Although the technical feasibility of the process has been demonstrated, an economical analysis may be necessary to evaluate the real possibility of application of this process on the sludge treatment to remove heavy metals.❷ According to the proposed bioleaching mechanism, the metals solubilisation is ascribed to the production of sulphuric acid by microorganisms, that are able to utilize a substrate containing reduced sulphur. For this reason, the process can be favourably applied either when the substrate contains a source of sulphur, or when it is possible to add ferrous sulphate obtained as a by-product.

Fly ashes can be considered as renewable secondary sources for the microbial recovery of heavy metals. Biohydrometallurgical processing of fly ash from municipal solid waste incineration resulted in significant recovery of Zn, Al, Cu, Ni and Cr using a mixed culture of iron and sulfur oxidizing bacteria. Here the process requires adaptation of the bacteria in CFA as bioleaching inhibits bacterial growth creating alkaline pH and dilution effect of sulfur particles.

Ishigaki et al. investigated the behavior and characteristics of metal leaching from municipal solid waste incineration (MSWI) fly ash among pure cultures of a sulfur-oxidizing bacterium (SOB) and an iron-oxidizing bacterium (IOB) and a mixed culture. It was observed that the IOB had a high metal-leaching ability, though its tolerability against the ash addition was low. The SOB might better tolerate an increase in ash addition than the IOB, though metal leaching ability of the SOB was limited. They argued that the mixed culture could compensate for these deficiencies, and high metal leachability was exhibited in the 1% ash culture, e g. 67% and 78% of leachabilities for Cu and Zn, respectively, and 100% for Cr and Cd. Characterization of metal leaching by the mixed culture revealed that the acidic and oxidizing condition had remained stable thorough the experimental period. They concluded that in the recovery of the valuable metals from MSWI fly ash, bioleaching using a mixed culture of SOB and IOB was a promising technology.

7.3 Metal Recovery from Mine Waste and Nuclear Waste

The acid mine drainage from mining activities at Falu copper mine in Sweden has caused low pH values and high metal levels in the nearby Falu river and it's surroundings. A long-term solution to this environmental problem is to treat the mine water overflow in such a way that a neutral pH and an essentially metal free aqueous effluent is obtained. The idea is to first oxidize the ferrous iron with bacteria, in order to make a selective ferric iron precipitation at a pH of around 3. One possible process option is to use the precipitated ferric hydroxide for the production of red pigment.[8] Thereafter, zinc will be recovered for recycling.

The ferrous iron oxidation was initially studied in laboratory scale using batch cultures of mesophilic (35°C), moderate thermophilic (45°C) and extreme thermophilic (65°C) microorganisms. The ferrous iron oxidation kinetics was determined with two different concentrations of ferrous iron. The moderate thermophilic culture did not grow well on ferrous iron only. Since the mesophilic and extreme thermophilic microorganisms showed approximately the same oxidation kinetics, the mesophilic bacteria was selected for further studies in pilot scale, due to their lower operating temperature which reduces the heating cost.

A pilot plant was erected at the mine site with three 500L reactors in series with a treatment capacity of up to approximately 500L/h. The reactors were filled with plastic bodies in order to support the formation of a permanent biofilm to avoid bacterial washout. The pilot plant has been operating continuously for several months and the influence of flow rate, ferrous concentration, pH and temperature was investigated. With a ferrous concentration of 3.5g/L, 35°C, pH 1.8 and a flow rate of 330L/h a ferrous iron oxidation rate of 750mg/(L·h)was achieved.

Microorganisms of three different temperature ranges were used. The mesophilic bacteria were maintained at 35°C and contained species like *T. ferrooxidans*, *Leptospirillum ferrooxidans* and *Thiobacillus thiooxidans*. The moderate thermophiles, operating at 45°C, were also a mixed culture containing non-identified ferrous oxidising bacteria and the sulphur oxidiser *Thiobacillus caldus*. The extreme thermophilic microorganisms were a culture of *Sulfolobus metallicus* and was maintained at 65°C. The bacteria were maintained at pH 1.6 on a medium containing sulphur and ferrous iron and before the start of the experiments, the bacteria were sub-cultured for three to five times on ferrous iron only.

A schematic flowchart of the biooxidation process is seen in Fig. 7.2. Mine water was pumped

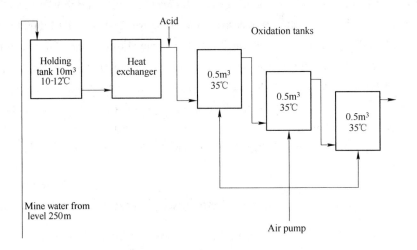

Fig. 7.2 Principal layout of the biooxidation plant at Falu mine

to a 10m³ holding tank from which the water was pumped at the desired flow rate through a heat exchanger, where the temperature was raised from 10-12°C to 35°C. The pH in the solution leaving the heat exchanger was adjusted to 1.6 with addition of sulphuric acid, before being fed into the bottom of the first oxidation tank. The oxidation unit consisted of three 500L tanks in series, where the overflow from one tank was fed into the bottom of the next tank in the series. A suspended biofilm carrier filled the first two oxidation tanks to 50% and the third tank to 40%. The biofilm carrier, Natrix™ (supplied by ANOX AB, Lund, Sweden) was made of high-density polyethylene and was designed to provide a large surface area for bacterial attachment. Air, enriched with 1% CO_2, was added in the bottom of the tanks through a rubber membrane. During the first 90 days, the air flow rate was maintained at 280L/min and was thereafter increased to approximately 600L/min. The airflow caused the biofilm carriers to move in the tank, and thereby also served the purpose of mixing. A fluid nutrient solution (Nutriol™ from Hydro Care, Landskrona, Sweden) was added at a rate of 360mL/h. The fluid nutrientsolution with a density of 1.27kg/L contained 22% total nitrogen in the form of urea,ammonia and nitrate and 2.5% phosphorus as phosphate.

During the 4-month period of the pilot plant operation, the ferrous iron content in the mine water varied irregularly in the range 0.75-4.4g/L, and the pH was between 1.5 and 3.1. The redox potential varied between 340mV and 430mV, depending on the ferrous iron concentration in the water that was pumped from the mine.

The ferrous iron oxidation experiments were started with a culture that had just completely oxidized a solution containing 10g/Lferrous iron in the 9K medium, therefore, the lag phase for all types of bacteria was short. The oxidation kinetics was followed by titration of ferrous iron. The decrease in ferrous concentration with time was linear and the results on the oxidation kinetics. It was found that the moderate thermophilic bacteria at 45°C were less efficient in oxidising the ferrous iron. As mentioned earlier, the stock cultures were grown on ferrous iron and elemental sulphur and before the experiments were started the bacteria were sub-cultured on ferrous iron for three to five times. It was found that for each sub-culture of the moderate thermophilic culture, the time needed to oxidise the 10g/L ferrous iron solution was longer. It is therefore concluded that these bacteria need sulphur or sulphides in order to grow and oxidize ferrous iron efficiently. This was not observed for the other types of microorganisms.

The mesophilic culture at 35°C and the extreme thermophilic culture at 65°C produced similar rates of ferrous oxidation. The rate of oxidation increased by three times at 35°C and by five times at 65°C with the increase in ferrous iron concentration from 2g/L to 17g/L. For the extreme thermophilic culture, it was found that the dependence of the amount of inoculum on the oxidation rate was minor, while the rate for the mesophilic culture was increased by 61%.

The percentage of the ferrous iron that was oxidised to the ferric form was calculated with the aid of the redox electrode calibration curve and is shown in Fig.7.3. As can be seen from this figure, a high degree of ferrous iron conversion was obtained at times of stable operating conditions. Based on the feed rate and the degree of ferrous conversion, the ferrous iron oxidation rate was calculated. The highest rates of oxidation, 700-750mg/L, were obtained during the period from day 55 to day 67 with a feed rate of 330L/h and a ferrous content of approximately 3g/L. It is believed that higher oxidation rates would have been obtained when the feed rate was increased to 400 L/h if pH was controlled successfully throughout the test period.

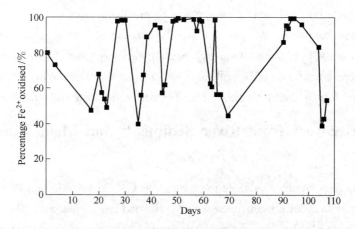

Fig. 7.3 Percentage of the ferrous iron oxidised during the campaign

During the campaign, there were problems with the heat exchanger on two occasions, on day 44 and on day 57, resulting in decrease in temperature down to 25°C. These problems were however overcome within 1-2 days on both occasions. The decrease in temperature resulted in an immediate decrease in the ferrous iron oxidation activity. When the temperature control was restored, the bacteria regained the activity quickly and high redox potential was recorded. Fast oxidation kinetics were also obtained at low temperatures. It might be possible, that if the bacteria had been allowed to adapt to lower temperatures for some time, good oxidation rates would also have been obtained at 25°C or even lower temperatures.

The oxidation kinetics of ferrous iron using a mesophilic culture at 35°C and an extreme thermophilic culture at 65°C in stirred suspended-cell batch reactors gave similar results. The moderate thermophilic culture used did not grow well on ferrous iron only. The suspended carrier process for bacterial oxidation of ferrous iron has been proven to be technically feasible. The process was easy to operate and during the pilot plant campaign, only some minor technical problems were encountered. The bacterial activity was easily restored after disturbances in control of pH and temperature. With mine water containing approximately 3.5g/L ferrous iron, a retention time of 3.6h is considered sufficient for complete conversion of the iron into the ferric form.

The major drawbacks for the process involving bacterial oxidation of mine water from Falu mine is the extremely high levels of ferrous iron. When the water level in the mine is held constant, the expected ferrous concentration is 6.7g/L. The high iron content requires a longer retention time for complete oxidation of the ferrous iron. The consumption of sulphuric acid in order to control pH was also relatively high.

Recently, it has been revealed that rare metals including vanadium, uranium and radionuclide can be mobilized by bacterial action. Soil contaminated with crude oil was tested in the presence of wide range of bacteria. The sorption capacity of vanadium of the biomass and bacterial tolerance to vanadium salt was investigated using the minimum inhibitory concentration (MIC) test. Organic acids and ligands produced by the bacteria made a change in the pH and chelation regime with a consequent mobilization of the metals. Some siderophore producing bacteria such as *Pseudomonas fluorescens*, *Shewanella putrefaciens* and *Pseudomonas stutzeri* have been found to mobilize uranium and other elements from uranium ores of shale mine tailings. Uranium can be mobilized from ore under aerobic conditions at neutral to alkaline conditions. John et al. have also mentioned that the uranium can be recovered from the industrial waste (e.g. nuclear industry) by using microorganism. Some other researchers have studied the microbial degradation of cement solidified radioactive wastes. When nutrients are available the microorganisms form a biofilm on the surface of cement solid waste and remain active.

7.4 Metal Recovery from River Sediments and Metal Finishing Waste Water

Efficient microbial dissolution of valuable metals such as Cu, Co, Ni, Mn and Fe has been achieved from ocean manganese nodules in the presence of pyrite and reducing agents. Heavy metal leaching from anaerobic sludge has also been reported using pyrite as an energy source. Some components of river sediments have been found to play a significant role in the bioleaching of metals from contaminated sediments. The solid content determines the size of the bioreactor and operational time.

It has been reported that iron and sulfur oxidizing bacteria solubilized more than 90% of Ni, Zn, Cu and Cr from river sediments at an optimum temperature of 37°C because at high temperature, pH changes prevent initiation of the indirect attachment mechanism. Changes in oxidation-reduction potential (ORP) also indicate the level of activity of sulfur oxidizing bacteria. The ORP at 37°C increased more rapidly than it did at the other temperatures (25°C and 55°C) (Fig. 7.4). The difference in ORP at 37°C became apparent after 7 days, and grew larger from then onwards. This indicates that there was more oxygen input and uptake to satisfy sulfur-oxidation by sulfur oxidizing bacteria in the sediment solution at 37°C. A third indicator of bioleaching efficiency is sulfate concentration. The amount of sulfides oxidized to sulfates at 37°C was markedly higher than at other temperatures (Fig. 7.5). After

18 days at 37°C, the accumulated concentration of sulfates was about 18,000mg/L whereas the accumulated concentrations at 25°C and 55°C were only about 7,000mg/L. This confirms that, of the three temperatures tested, the sulfide-oxidizing activity of sulfur-oxidizing bacteria in sediment is significant only at 37°C.

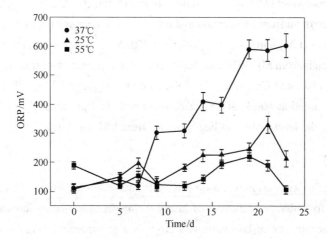

Fig. 7.4　Effect of temperature on ORP during bioleaching

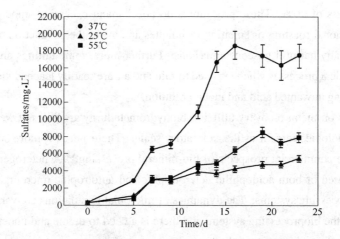

Fig. 7.5　Effect of temperature on sulfates during bioleaching

Diaz et al. evaluated nickel and cobalt recoveries from tailings of Caron technology process using sulphuric acid produced by Acidithio-bacillus thiooxidans cultures under different conditions. They reported that high nickel and cobalt recoveries (about 60% for cobalt and 85%-100% for nickel) had reached after 13 days when low pulp densities (1% and 2.5%) of laterite tailings were used. However, they observed negligible metal recoveries when higher pulp densities were used.

Ndlovu et al. treated nickel laterites with chemolithotrophic microorganisms using H_2SO_4 acid, citric acid and acidified $Fe_2(SO_4)_3$. Results showed that H_2SO_4 acid performed better, in terms of nickel recovery, than citric acid or acidified $Fe_2(SO_4)_3$. Mixed cultures of *A. ferrooxidans*, *Acidithiobacillus caldus* and *L. ferrooxidans* were used in the presence of elemental sulphur and FeS_2 as energy sources in that bacterial leaching works. The sulphur substrate exhibited better effects in terms of bacterial growth, acidification and nickel recovery than the FeS_2 substrate. They observed maximum nickel recovery of 79.8% with the optimum conditions (an initial pH of 2.0, 63μm particle size and 2.6% pulp density).

Metal finishing wastewater (MF-WW) contained heavy metals (Cr, Cu, Pd and Zn) and Sthiannopkao and Sreesai recovered them from by sorption and precipitation mechanisms using pulp and paper industrial wastes, lime mud (LM) and recovery boiler ash (RBA). LM demonstrated better performance in recovering heavy metals from MF-WW than RBA. At a reaction time of 45 min, the maximum recovery efficiencies for Cr (93%) and Cu (99%) were obtained at 110g/L of LM, while for Pb (96%) and Zn (99%) they were obtained at 80g/L of LM. Treatment with LM produced higher sludge volume than with RBA. However, the leach ability of heavy metals from LM was lower.

7.5 Summary

Today, biohydrometallurgy is no longer a promising future technology but an actual and effective economical alternative for enhanced recovery of heavy metals from low-grade ores and wastes. There is an ever-increasing acceptance of biohydrometallurgy as a commercially viable technology due to its distinct technical and cost advantages over traditional physicochemical techniques. Almost without exception, microbial extraction procedures are more environment friendly, while giving high extraction yields in excess of 90%. These procedures do not demand excessive amounts of energy when compared to traditional roasting or smelting techniques and avoid the production of sulfur dioxide or other environmentally harmful gaseous emissions. Further-more, mine tailings and wastes produced from physicochemical processes when exposed to rain and air are readily leached via natural biological processes, producing unwanted acid and metal pollution.

Research efforts on metal recovery utilizing biohydrometallurgy are still not well advanced and in many cases, technologies remain at the laboratory scale. There needs a more specific research effort to advance the commercial prospects of bio-mineral processing.❹ A heterogeneous and complex microflora, composed of both acidophilic heterotrophic and autotrophic microorganisms, exists with commercial bioprocessing systems. The dynamics of microbial populations are variable both spatially and temporally in the bioprocessing system, and there is a need to define and understand the potential interactions among the components of the microflora.

Currently, industrial biomining processes have utilized bacteria that grow optimally from ambient temperatures to 50°C. There is considerable potential to develop processes that can operate using thermophilic microbes at temperatures of 80°C and above. This process will make it possible to recover metals from a wider range of minerals economically.

Questions

1. What is the difference between biohydrometallurgy and bioleaching?

2. How to utiliz the *Thiobacillus ferrooxidans* to recover the heavy metals?

3. Why was the ferrous iron oxidation studied at three different temperatures, 35°C, 45°C and 65°C, respectively?

4. How does the temperature affect the ORP value during bioleaching?

5. How do you think about utilizing biohydrometallurgy to recover the metals from industrial wastes?

Note

注　释

❶ 生物冶金过程是通过金属与微生物之间的相互作用将难溶金属硫化物转变成可溶的金属硫酸盐。

❷ 尽管生物浸出过程的技术可行性已被认可，但利用该技术处理污泥并提取重金属的工艺仍需要进行经济效益评价。

❸ 首先使用细菌氧化亚铁离子，在 pH 值为 3 左右生成三价铁沉淀。生成的氢氧化铁沉淀可以用来生产红色颜料。

❹ 利用生物冶金回收金属的研究工作依然进步不大，很多技术还停留在实验室规模上。需要进一步加强研究，促进商用前景的开发工作。

8 Biohydrometallurgical Recovery of Value Metals from Secondary Sources

本章要点 电子废弃物、废旧电池、石油化工催化剂等二次资源中含有多种有价金属，传统上主要采用酸法、电化学法和火法进行回收处理。本章简要介绍生物冶金技术在处理这些二次资源中所使用的细菌种类和不同的工艺方法，以及所取得的效果。

8.1 Introduction

World production of electronic, battery wastes and other related solid wastes is constantly increasing, both in industrialized and in developing countries. Governments along with the public have paid high attention to the treatment of such wastes not only due to their hazardous nature but also to the valuable metals. Generally, acid leaching and electrowinning processes are applied to recover the metals from these wastes. Microbiological processes can also be utilized to mobilize metals from these waste materials.

Metals including Al, Ni, Pb and Zn have been reported to be extracted successfully by biohydrometallurgical process. Bacteria and fungi can produce inorganic and organic acids and cause mobilization of metals. In the recovery of value metals from electronic wastes, biohydrometallurgical process showed much better performance over pyrometallurgical process.

In developing countries such as China, bioleaching techniques have been applied to extract toxic metals from hazardous spent batteries. For example, by using a two-stage process, thiobacilli which was grown on sewage sludge and elemental sulfur was used to generate acid in involving a bioreactor and a settling tank. Then, the formed metabolites received by a leaching reactor were used for metal solubilization. This system was able to leach metals effectively from Ni–Cd spent batteries.

This chapter will be specially dealt with the biohydrometallurgical technologies recovering electronic wastes, spent batteries and petroleum refinery catalyst as well.

8.2 Metal Recovery from Electronic Wastes

Various hydrometallurgical processing techniques including cyanide leaching, halide leaching, thiourea leaching, and thiosulfate have been reported to leach precious metals from electronic wastes.

Brandl et al. showed that a bacterial leaching process or the mobilization of metals from fine-grained electronic waste. *Penicillium simplicissimum* and *Aspergillus niger* were used in a two-step bioleaching process. In the first stage, biomass was produced in the absence of electronic scrap. Subsequently, electronic scrap was added in different concentrations and the cultures were incubated for an additional time period.❶

Initial experiments demonstrated that at a concentration of >10g/L of scrap in the medium, microbial growth was inhibited. However, after a prolonged adaptation time of at least 6 weeks, fungi grew also at concentrations of 100g 10g/L. Inhibiting compounds were not identified. During growth on sucrose, various organic acids citrate, gluconate, oxalate were formed. After an incubation period of 21 days, *A. niger* formed 3mmol/L. oxalate and 180mmol/L citrate, whereas *P. simplicissimum* formed 5mmol/L oxalate and 20mmol/L citrate in the same period. Fig. 8.1 shows the metal mobilization percents by leaching of from electronic scrap after 21 days at 30°C. In general, *P. simplicissimum* was able to mobilize more metals as compared to *A. niger* under the same conditions. *A. niger* has a certain preference to mobilize copper. Both fungal strains were able to mobilize Cu and Sn by 65%, and Al, Ni, Pb, and Zn by more than 95%.

The results also show that the two-step process seems appropriate to increase leaching efficiency. The process direct growth of organisms in the presence of electronic scrap is poorly suited and not advisable. For a commercially, two-step process is appropriate for an industrial application. In the first step, organisms will be grown in the absence of electronic scrap followed by the second step where the formed metabolites will be used for metal solubilization. Furthermore, leached and recovered metals might be recycled and re-used as raw materials by metal-manufacturing industries.

Creamer et al. reported the studies on recovery of precious metal recovery from electronic scrap leachates by *Desulfovibrio desulfuricans*. Samples of electronic scrap were shredded printed circuit board fragments from a commercial plant and granulated electronic scrap with the iron and aluminum content previously reduced. The approximate contents of the metals of interest were: Cu: 20%-25wt%; Au: 60-113 ppm; Pd: up to 30mg/L. Au was precipitated extracellularly by a different mechanism from the biodeposition of Pd. The presence of Cu^{2+} (−2000mg/L) in the leachate inhibited the hydrogenase-mediated removal of Pd(II) but pre-palladisation of the cells in the absence of added Cu^{2+} facilitated removal of Pd(II) from the leachate and more than 95% of the Pd(II) was removed autocatalytically from a test solution supplemented with Cu(II) and Pd(II). Metal recovery was demonstrated in a gas-lift electrobioreactor with electrochemically generated hydrogen, followed by precipitation of recovered metal under gravity.

Faramarzi et al. reported their preliminary investigation on the feasibility of recovery of gold from printed circuit boards by bioleaching process. First, different cyanogenic bacterial strains (*Chromobacterium violaceum, Pseudomonas fluorescens, Bacillus megaterium*) were cultivated under cyanide-forming conditions in the presence of nickel powder. All microorganisms were able to form water-soluble metal cyanides, however, with different efficiencies. *C. violaceum* shows the best performs. Approximately 4mmol/L of free cyanide was detected during growth of *C. violaceum*. Generally, cyanide concentration always decreased after reaching a maximum. This is probably due to cyanide consumption by cyanide-consuming (cyanicidic) compounds or to cyanide degradation. Only very small cyanide

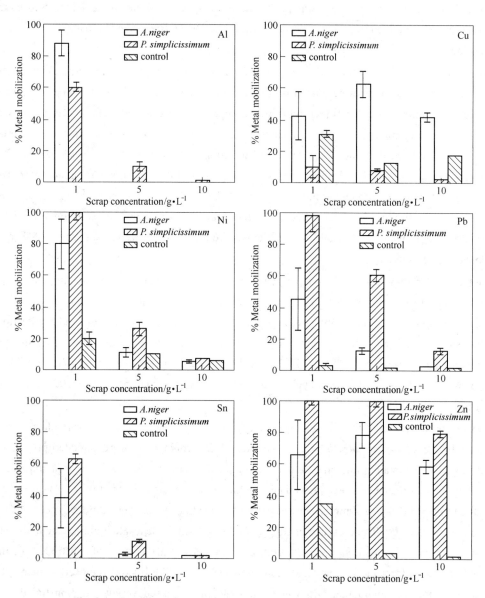

Fig. 8.1 Leaching of metals from electronic scrap after 21 days at 30°C on different concentrations of electronic scrap

concentrations were measured in the gas phase of closed vessels. *C. violaceum* was able to mobilize nickel. Cyanide-complexed nickel was already detected 4h after inoculation. Approximately 9% of the initial nickel powder added was recovered as tetracyanonickelate. The pH of the culture increased during incubation from 7.2 to approximately 9.

Then, Gold containing pieces (5mm × 10mm) were used for growth experiments. Gold containing pieces were obtained by cutting printed circuit boards followed by manual sorting. Each piece contained approximately 10mg of gold.❷ Using *C. violaceum*, as shown in Fig. 8.2, it is demonstrated that gold can be microbially solubilized from printed circuit boards. The maximum dicyanoaurate [$Au(CN)_2^-$]

measured corresponds to a 14.9% dissolution of the initially gold added. In conclusion, the results represent a novel type of microbial metal mobilization based on the ability of certain microbes to form HCN. The findings might have the potential for amicrobially based industrial application regarding the treatment of metal-containing solids or the biological remediation of metal-polluted soils since metal cyanides can be separated by chromatographic means and easily be recovered by sorption onto activated carbon. However, cyanogenic microbes have to be comprehensively evaluated to fully exploit the potential to form water-soluble metal complexes from solid metal-containing solids. Growth and cyanide formation have to be optimized as well as metal mobilizing efficiencies. In particular, the presence of cyanide-consuming compounds in the growth medium originating from the solids treated has to be addressed.

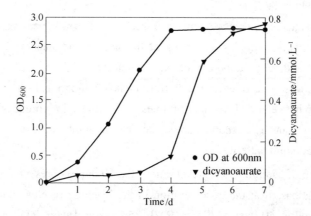

Fig. 8.2 Growth and dicyanoaurate formation by *C. violaceum* in LB medium supplemented with glycine and shredded pieces of printed circuit boards originating from electronic scrap

In industrially practiced or demonstrated, many potential lixiviant systems that can be exploited for extraction of precious metals from electronic wastes is well-established. Table 8.1 summarizes main features of these lixiviant systems for Au. Cyanide leaching is very effective for leaching gold, economically extract gold with grades as low as 1-3g/t Au and it is well established process. Thiosulfate leaching has the advantage of low cost, environmentally friendly, fast leaching rate, but its chemical stability is low. Thiourea leaching has fast leaching rate, but it also has low chemical stability and contain potential carcinogen. Halide leaching is relatively healthy and safe and has high chemical stability, but it is at the low developmental now due to apply more difficultly. So these methods have the advantage and disadvantage on the different aspects of economics, applicability and toxicity.

Table 8.1 Summary of features of potential leaching systems for extraction of gold from electronic wastes

Lixiviant	Reagents	Chemistry
Cyanide	CN^-, $Air(O_2)$	$4Au+8CN^-+O_2+2H_2O=4Au(CN)_2^-+4OH^-$
Thiosulfate	$S_2O_3^{2-}$, NH_3, Cu^{2+}	$4Au+8S_2O_3^{2-}+O_2+2H_2O=4Au(S_2O_3)_2^{3-}+4OH^-$
Thiourea	$CSN(NH_2)_2$, Fe^{3+}	$4Au+4CS(NH_2)_2+2Fe^{3+}=2Au(CS(NH_2)_2)_2^+ +2Fe^{2+}$
Halide	Cl^-/Cl_2, Br^-/Br_2, I^-/I_2	$2Au+11HCl+3HNO_3=2HAuCl_4+3NOCl+6H_2O$

8.3 Metal Recovery from Battery Wastes

According to the report prepared by the State Environmental Protection Administration of China (SEPA), the quantity of dry batteries produced and consumed has increased rapidly in recent years in China. In one year, 14.1 billion batteries were produced and 8.37 billion batteries were consumed which was reported in SEPA of year 2000 (SEPA, 2000) Some batteries such as nickel-cadmium batteries are classified as hazardous waste because nickel and cadmium are heavy metals and suspected carcinogens. However, only 1%-2% of discarded batteries are recovered in China due to a lack of relevant regulations. Therefore, it is necessary to find an economic and environment friendly process to recycle dry batteries. Some researches were carried out to investigate the metal recovery from battery wastes by the method of biohydrometallurgy.

Zhu et al. recovered anodic and cathodic material from three pairs of spent nickel-cadmium batteries with the same brand. Three samples consisting of the same component were made. Each sample contained 9.3g anodic and 9.1g cathodic materials, which include 2.34g cadmium and 6.23g nickel. *Indigenous thiobacilli* was used to leach metals from nickel-cadmium batteries. The first step was to clarify the relationship between RTB (residence times of the sludge in bioreactor) and the pH value of the settling tank supernatant. The second step was to determine the effect of leaching metal from the electrodes of nickel-cadmium batteries.

The total dissolution of cadmium and nickel during the 50 days is presented in Table 8.2. According to the calculations based on the concentration in the leaching reactor effluent, release of all of the cadmium only requires 40 days at 5 days RTB, 35 days at 4 days RTB, and 50 days at 3 days RTB. However, the total nickel recovered during the 50 days was only 66.1%, 75.6% and 40.8% for the RTB of 5 days, 4 days, and 3 days, respectively. A higher metal concentration in effluent occurred with a longer RTB during the leaching process (the metal concentration at 5 days RTB was more three times than that at 3 days RTB). However, due to the different total amount of solution used in each case, the metal leaching efficiency did not change based on the RTB only. The metal leaching is most effective at a residence time of 4 days. The results showed that the system was valid to leach metals from nickel-cadmium batteries, and that the sludge drained from the bottom of the settling tank could satisfy the requirements of environmental protection agencies regarding agricultural use.

Table 8.2 Summary of the results of total metals dissolution at different RTB

RTB day	Total metals dissolved/mg		Contribution of sludge/mg		Total solution/L
	Cd	Ni	Cd	Ni	
5	2341.4	3467.8	0.67	19.36	10.0
4	2341.4	3974.4	0.81	22.70	12.5
3	2340.1	2162.0	0.94	26.09	16.7

Cerruti et al. studied the indirect dissolution process of spent nickel-cadmium batteries. Two percolators system was used for the dissolution of a spent nickel–cadmium battery. The first percolator was a sulphuric acid bioreactor with *Thiobacillus ferrooxidans* immobilised on elemental sulphur. The acidic medium produced in the first percolator was pumped to the second percolation column which contained a whole spent battery previously broken. The battery was placed in the percolation second column and media produced in the *T. ferrooxidans* bioreactor (pH about 1.0) were added. The media were previously filtered through medium fast speed filter-paper to eliminate elemental sulphur but not colloidal sulphur or suspended cells. The liquid phase in the column was pumped to the top of the percolation column by air pressure. When the pH was too high (higher than 2.5) or metal concentrations in the solution did not change, all the medium was replaced by fresh medium in order to continue metal dissolution.

After 93 days of the indirect dissolution. 100% of cadmium, 96.5% of nickel and 95.0% of iron were recovered. Moreover, recoveries higher than 90.0% were reached when anodic and cathodic materials were directly added to *Thiobacillus ferrooxidans* cultures with sulphur as the sole energy source. The results presented show an economic and effective method which could be considered the first step to recycle spent and discarded batteries preventing one of the many problems of environmental pollution.

Velgosová et al. researched the influence of H_2SO_4 and ferric iron on Cd bioleaching from spent Ni-Cd batteries. The main aim of this study was to understand which from the bioleaching products (sulphuric acid or ferric sulphate) play a main role in the bioleaching process of Cd recovery. The best leaching efficiency of Cd (100%) was reached by bioleaching and also by leaching in $Fe_2(SO_4)_3$ solution (Fig. 8.3 and Fig. 8.4). It can be seen that the beginning of the leaching curves of both electrode material starts at about 40% of yield in the first day of bioleaching which corresponds to the dissolution of cadmium hydroxides in the first hours of the leaching process. The bioleaching rate is the highest during the early 7 days and during that time majority of Cd is dissolved from anodic material and on the following days the process rate is slowing down. However, in the case of cathodic material Cd bioleaching is gradually increasing until the 28th day. As there is the different chemical composition of both electrode materials and the anodic material mainly consists of cadmium hydroxides being much more soluble than metallic cadmium, the attack of anodic material is faster that corresponding to cathodic material. It is also necessary to take into account the toxicity of a higher amount of Cd on bacteria resulting in the inhibition of

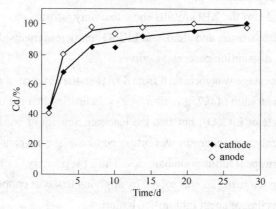

Fig. 8.3   Cd leaching yields in 9 K solution with *Acidithiobacillus ferrooxidans* bacteria

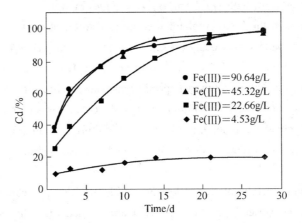

Fig. 8.4 Cd leaching yields of cathode powder in the solutions with different Fe concentrations

bioleaching process. Ferric iron as one of the main leaching agents plays the important role not only as the strong oxidative agents but also as a factor that can maintain the low pH throughout the experimental period, the environment needed for the Cd release, due to its hydrolysis. The use of H_2SO_4 solution resulted in the lowest efficiency of Cd leach ability, which is under 40%, maybe the presence of hydroxides in electrode materials caused neutralization of the leaching solution and inhibition of Cd leaching.

Zeng et al. investigated the bioleaching process of recycling cobalt from spent lithium-ion batteries (mainly $LiCoO_2$). 43.1% of cobalt dissolution was obtained after 10 days by *Acidithiobacillus ferrooxidans* bioleaching. Then they developed a copper-catalyzed bioleaching process of *Acidithiobacillus ferrooxidans*. Fig.8.5 shows the particle size and morphology changes of $LiCoO_2$ after leaching in the copper-catalyzed solution of different concentration. Clear differences of the residues morphologies in the presence of different copper ion concentrations were observed. When the experiment was performed in the absence of copper ion and with copper concentration of 0.001g/L, residues surface were coated with a solid product layer, a relative larger particle size of the residues was observed.[8] However, when the experiment was performed at Cu^{2+} concentration of 0.75g/L, a extreme exiguous particle was observed. Associated with the EDX analysis results that little Co contents in the presence of Cu^{2+} concentration of 0.75g/L, and also the XRD results show that only jarosite was existed in the residues. So the observed particles were jarosite, and almost no $LiCoO_2$ particles remained. These results are also in good agreement with the dissolution percentage curves.

Cobalt dissolution percentage was increased from 43.1% in the absence of copper ion within 10 days to 99.9% at copper concentration of 0.75g/L after 6 days, indicating that the copper ions could enhance not only the oxidation rate of $LiCoO_2$, but also the leaching amount of cobalt from spent lithium-ion batteries. A probable catalytic mechanism was proposed to explain the catalytic effect of copper ions on bioleaching of Co from spent lithium-ion batteries. Thus, the primary leaching efficiency problem of Co recovery from lithium ion batteries was solved by applying low cost copper ions as catalysis, which is very important for recycling of spent lithium-ion batteries.

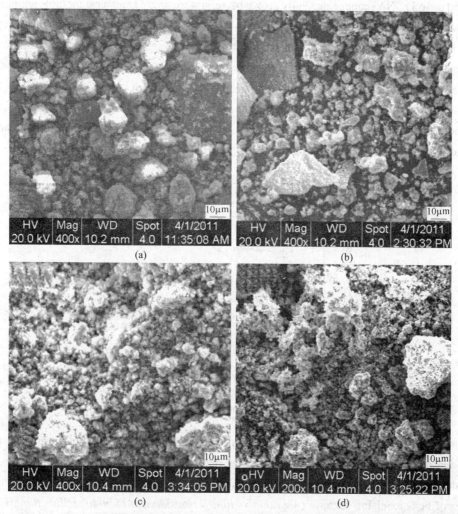

Fig. 8.5 SEM micrographs of leaching residues

(a) blank; (b) Cu^{2+} concentration of 0.01g/L; (c) Cu^{2+} concentration of 0.75g/L; (d) Cu^{2+} concentration of 10g/L

8.4 Metal Recovery from Spent Petroleum Refinery Catalyst

Spent petroleum catalyst contains various valuable metals that have great potential for re-use. Pradhan et al. investigated the bioleaching and chemical leaching for dissolving Ni, V, and Mo from spent petroleum refinery catalyst. The spent catalyst for the leaching experiments was collected from the Korean Petroleum Company. Three different size fractions of spent catalyst powder, <45 μm, 106-45 μm, and 212-106 μm were used. The metal contents of the de-oiled spent catalyst powder were (wt.%): Al, 19.5%; S, 11.5%; Ni, 2.0%; V, 9.0%; Mo, 1.4%; and Fe, 0.3%. Two step leaching experiments were carried out to extract nickel, vanadium, and molybdenum present in spent refinery catalyst. A bioleaching process was applied in the first step.

The bacteria used in the bioleaching experiment were cultured from a water sample collected from the Dalsung mine area in South Korea. After several subculturing processes, the biomass was collected and subjected to 16S rDNA sequencing. Fully grown bacteria were filtered and used for bioleaching. During the bioleaching process, 97%, 92%, and 53% Ni, V, and Mo, respectively, leached out under optimum conditions. As only 53% of the Mo leached out during bioleaching, the recovery of residual Mo was the main objective in this step. Further leaching was considered to recover the residual Mo using different chemical reagents. Second-step leaching was conducted using H_2SO_4 Na_2CO_3, and $(NH_4)_2CO_3$, to extract the residual Mo. The overall leaching efficiency of Mo, obtained by combining both bioleaching and chemical washing, is shown in Fig. 8.6. $(NH_4)_2CO_3$ showed a better result, so further leaching of the residue from the first step was considered using $(NH_4)_2CO_3$.

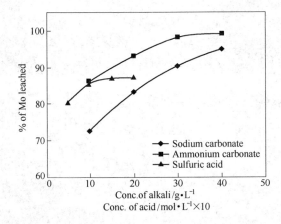

Fig. 8.6 Overall leaching of Mo with the combination of bioleaching and chemical leaching

Beolchini et al. report a bioleaching study aimed at recovering metals from hazardous spent hydroprocessing catalysts. The exhaust catalyst was rich in nickel (4.5mg/g), vanadium (9.4mg/g) and molybdenum (4.4mg/g). A biotechnological environmentally friendly strategy, involving bioleaching abilities of Fe/S oxidizing bacteria (*Acidithiobacillus ferrooxidans*, *Acidithiobacillus thiooxidans* and *Leptospirillum ferrooxidans*), has been applied on Italian refineries spent hydroprocessing catalysts.

Investigated factors were elemental sulphur addition, ferrous iron addition and actions contrasting a possible metal toxicity (either adding powdered activated charcoal or simulating a cross current process by means of periodical filtration). Fig. 8.7 shows nickel, vanadium and molybdenum extraction yields observed in all treatments after 21 days bioleaching. These values are important when designing the downstream operations for metals purification and recovery. It can be observed that the highest extraction yields for Ni and V (83%±4% and 90%±5%, respectively) were achieved in treatments 3 and 4, which are those in the presence of iron with no application of any strategy aimed at inhibiting metal toxicity. Furthermore, the effect of ferrous iron on the biological activity of the microbial consortia is demonstrated by the significant difference between inoculated flasks (biological treatments) and not inoculated ones (chemical controls). An analysis of the observed extraction yields in treatments 9-12 suggests that the periodical filtration of the liquor leach, aimed at inhibiting metal toxicity, had, on the other hand, a negative effect on metal mobilisation: In fact, the highest values experimentally observed for Ni and V extraction yields were not higher than 40% for both metals. This may suggest that filtration (and re-suspension in fresh medium) removes compounds which may have a key role in metal dissolution. Activated carbon did not favour metal leaching as expected: in fact, no significant differences are apparent between treatments 1-4 and 5-8. As concerns molybdenum dissolution, the observed values for Mo extraction yields were not as high as Ni and V ones. The highest values were around 30%-40% for treatments where no strategy inhibiting metal toxicity was applied and around 5%-10% when periodically filtering the liquor leach. The so low Mo extraction yields observed in treatments 5 and 6 were supposed to be due to activated carbon adsorption of molybdate ions. In contrast with its effect on nickel and vanadium dissolution, iron did not seem to have a positive effect on molybdenum recovery: Mo extraction yields seem not to be enhanced by iron, even if a significant difference between biological treatments and chemical controls is remarkable also for molybdenum.❹

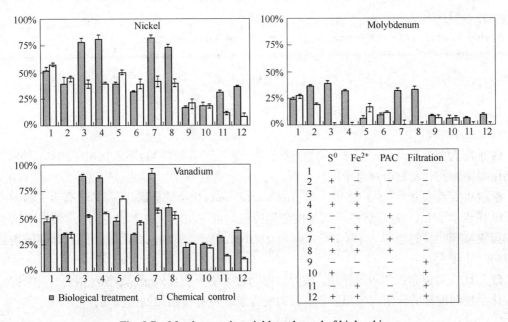

Fig. 8.7 Metal extraction yields at the end of bioleaching

8.5 Summary

Recycling nonferrous metals from secondary sources (for instances, electronic waste, battery wastes et al.) is an important subject not only from the point of waste treatment but also from the recovery of valuable metals. The value distribution for electronic scrap samples shows that for cell phones, calculators, and printed circuit board scraps, the precious metals make up more than 40%-70%. Spent Ni–Cd batteries contain a high amount of value metals (51%-67%), while in nature the content of cadmium occurs as a part of other minerals is only 0.02%-0.6%. This indicates that recycling of wastes is from the recovery of precious metals.

Compared with the conventional methods (pyrometallurgy and chemical hydrometallurgy), biohydrometallurgy process offers a number of advantages including low operating costs, minimization of the volume of chemical and high efficiency in leaching rate for extract value metals from many secondary sources.

Questions

1. What is the effect of bacteria and fungi during the biohydrometallurgical processing?
2. What is the two-step process that increases leaching efficiency for an industrial application?
3. What is the difference between the first step and the second step during leaching metals from nickel–cadmium batteries?
4. Why is a second step of residue chemical leaching needed for leaching Mo from spent petroleum refinery catalyst?
5. How to treat the spent hydroprocessing catalysts in Italian refineries?

Note
注　释

❶ 使用真菌进行两步法浸出，例如，第一步，单独培养微生物(不添加电子废料)，第二步，利用微生物的代谢物溶解电子废料金属。

❷ 浸出实验采用手工从电路板上取下的 5mm×10mm 含金的金属片，每片含有金大约 10mg。

❸ 图表示了 $LiCoO_2$ 颗粒在不同浓度的铜催化溶液中浸泡后的形貌变化，从图中可以看出残留物具有明显的差别。当铜离子浓度为 0.001g/L 时，残留物的表面被一层固体膜覆盖且残留物尺寸较大。

❹ 对比铁对镍和钒的溶解作用，铁对钼的回收无明显积极效果；尽管对钼而言，生物处理与化学控制过程的作用有明显差别，但钼的提取率没有被添加铁而提高。

Appendix
有色金属生物冶金精选专业词汇英汉对照表

A

ablation, n. 消融，切除
absence of 缺乏，缺少
absolute level 绝对水平；绝对位准
absorption, n. 吸收
accelerate, vi. 加速，促进；vt. 使……加快
acceptor, n. 接受者，接受器
accompany, vt. 陪伴，伴随
according to 根据
account for 对……做出解释
accumulate, vi. 累积；积聚
accumulation, n. 积聚，累积
accurately, adv. 精确地
acid, n. 酸；adj. 酸的
acidic, adj. 酸的，酸性的
acid condition 酸性条件
acid consumption 酸耗量
acid leaching [冶] 酸浸
acid level 酸度
acid resident 耐酸的
acid residue 酸性残渣
acid solution 酸溶液
acid-consuming 耗酸的
acid-generating 产生酸的
acidianus, n. [生] 酸菌属
acidic ferric lixiviant 酸性铁浸出剂
acidification, n. [化] 酸化，成酸性，使……发酸
acidified sulfate 硫酸酸化
acidiphilium, n. [生] 嗜酸菌属
acidithiobacillus caldus 硫杆菌
acidithiobacillus ferrooxidans 嗜酸氧化亚铁硫杆菌
acidithiobacillus thiooxidans 嗜酸氧化硫硫杆菌
acidity, n. 酸度，酸性
acidophile, n. 嗜酸细胞
acidophilic, adj. [生] 嗜酸的，耐酸的
acidophilic autotroph 嗜酸性自养生物

acid-producing, adj. 生酸的，产酸的
acid-solubility, n. [化] 酸溶性
actinomyces, n. 放线菌
activated carbon 活性炭
activating agent [助剂] 活化剂
activity, n. 活动，活跃，[物化] 活性
adapt, vi. 适应
adaptation, n. 适应，改编
adhere, v. 坚持，依附
adherence, n. 坚持，依附
adhesion, n. 黏附，附着力
adsorption, n. [物] 吸附 (作用)
adulterant, n. 掺杂物；adj. 掺杂用的
aeration, n. 曝气，通气
aerobe, n. [微] 需氧菌，需氧生物
aerobic, adj. 需氧的，好氧的
aerobic formers 好氧菌，需氧菌
aerosol, n. [物化] 气溶胶；adj. 喷雾的
agglomerate, n. 团块；adj. 凝聚的
agglomerating, adj. 附聚的；凝结的
agglomeration, n. 凝聚，团聚
agitator, n. 搅拌机
agricultural, adj. 农业的；农艺的
air cooling 空冷
air delivery 排气量
air demand 供气量
air injection system 空气注射系统
air ventilation system 空气净化系统
airlift reactor 气升式反应器
a large number of 大量的，很多
alkaline, adj. 碱性的，碱的
alter, vt. 改变，更改
alternative, adj. 供选择的，选择性的；n. 供替代的选择
ambient temperature 环境温度
ambient, n. 周围环境；adj. 外界的

amenability, n. 顺从，服从的义务
ammonia, n. 氨，氨气
ammonium, n. 氨气，氨盐基，铵
ammonium nitrogen 氨态氮
amount, vi. 总计，合计；n. 数量，总数
anaerobe, n. [微] 厌氧性生物，厌氧菌
anaerobic adj. [微] 厌氧的，[微] 厌气的
anaerobic bacteria [微]厌氧细菌
analysis, n. 分析
analyze, vt. 对……进行分析
anchor, vt. 抛锚，使固定；n. 锚
anerobic, adj. 厌氧的
anion, n. 阴离子
anodic, adj. 阳极的
anodic current 阳极电流
antibody, n. [免疫] 抗体
antimony, n. [化] 锑
antioxidant, n. 抗氧化剂，硬化防止剂
appendage, n. 附加物，下属
application, n. 应用
apply to 应用于，适用于
appropriate, adj. 适当的
approximately, adv. 大约，近似地
aqueous, adj. 水的，水般的
aqueous phase 液相，[冶] 水相
aqueous raffinate 水相萃残液
aqueous solution [化] 水溶液
aromatic kerosene 芳香煤油
Arrhenius equation [物] 阿伦尼乌斯方程
arsenic, n. [化] 砷；adj. 砷的，含砷的
arsenopyrite, n. [矿] 毒砂，[矿] 砷黄铁矿
as a result of 因此，由于
a series of 一系列的
aseptic condition 无菌条件
aspect, n. 方面
aspergillus niger 黑曲霉
assist, vt. 促进；n. 帮助
associated with 与……相关
assume, vt. 假定，承担
assumption, n. 假定，设想
at the expense of 以……为代价
atalyst, n. 催化剂，触媒

atmospheric, adj. 大气的，大气层的
attachment, n. 附件，附着，连接物
automatic crane 自动化起重机
autotroph, n. 自养生物
autotrophic bacteria 自养细菌
autotrophic, adj. [生]自养的，无机营养的
availability, n. 可用性，有效性
available, adj. 有效的，可获得的

B

bacillus, n. 杆菌，细菌
bacillus megaterium 巨大芽孢杆菌
backfill, vt. 回填，装填；n. 回填
bacteria, n. 细菌
bacterial activity 细菌活性
bacterial attachment 细菌附着
bacterial cell 细菌细胞
bacterial culture 细菌培养
bacterial growth rate 细菌生长速率
bacterial leach rate 细菌浸出率
bacterial leach 细菌浸出
bacterial oxidation reaction 细菌氧化反应
bacterial population 微生物丛
bactericide, n. [药] 杀菌剂
bacteriosmelt, n. [微] 细菌冶金
bacterium, n. [微] 细菌；杆菌属
base metal 基本金属
basicity, n. 碱性，碱度
batch reactor 间歇式反应器
batch, n. 一批，一炉；vt. 分批处理
battery, n. 电池，蓄电池
bauxite, n. [矿] 铝土矿，铝矾土
be accompanied by, prep. 伴随有
be available for 有效，对……有用
be bound to 一定会，束缚于
be caused by 由……引起
be composed of 组成……
be concerned with 涉及……
be dependent on 依赖，取决于
be divided into 被分成
be eliminated 被淘汰
be equal to 等于

be found 被发现
be identical to 与……相同
be related to 与……有关
be responsible for 对……负责，是……的原因
be similar to 与……相似
be thought to 认为
behavior, n. 行为
bench-scale leaching 规模型浸出
beneficiation, n. 选矿，富集
bioaccumulation, n. [化] 生物富集作用
bioanalysis, n. 生物分析法
biochemical, adj. 生物化学的
biodeposition 生物沉积作用
biofilm, n. 生物膜，菌膜
biohydrometallurgy, n. [化] 生物湿法冶金
bioleaching, n. [化] 生物浸出，细菌浸出
biological oxidation 生物氧化
biomass, n. [生态] 生物量
biooxidation 生物氧化，细菌氧化
bioprocess, n. 生物过程，生物处理
bioreactor, n. [细胞] 生物反应器
bioremediation, n. 生物降解，生物修复
blackening, n. 变黑；v. 变黑，诋毁
blast, vt. 爆炸；n. 爆炸
bonding orbital 成键轨道
bound, n. 范围，跳跃；vt. 束缚
boundary, n. 分界线，边界
boundary layer [数] 边界层
brass, n. 黄铜
brass scrap 废黄铜
build up 增进，加强
by the end of 在……之前
by-product, n. 副产品，意外收获

C

cadmium, n. [化] 镉
cadmium hydroxides [无化] 氢氧化镉
calcite, n. [矿] 方解石
calcium, n. 钙
calcium arsenite 亚砷酸钙
calibration curve [计量] 校正曲线，[自] 校准曲线
calomel electrode 甘汞电极

capsule, n. 胶囊；vt. 压缩；adj. 压缩的
carbon dioxide 二氧化碳
carbon dioxide concentration 二氧化碳浓度
carbonate, n. 碳酸盐
carbonate mineral 碳酸盐矿物
carcinogen, n. 致癌物质
carry out 实现，执行
catalyst, n. [物化] 催化剂
cathodic, adj. 阴极的，负极的
cathodic reaction 阴极反应
cation, n. [化] 阳离子，正离子
chalcocite, n. [矿] 辉铜矿
chalcopyrite, n. [矿] 黄铜矿
character, n. 品质，特性
charge, n. 电荷
charged species 带电粒种
charge-transfer coefficient 电荷转移系数
chelation, n. [化] 螯合作用
chemical breakdown 化学分解
chemical leaching 化学浸出
chemolithotrophic 无机化能营养的
chemostat, n. (培养微生物的) 恒化器
chloride, n. 氯化物
chromatographic, adj. 色析法的，层离法的
chromatography, n. 色谱分析法
chromium, n. [化] 铬
chromobacterium violaceum 紫色色杆菌
chromosomal, adj. 染色体的
chronicle, n. 编年史，记录；vt. 记录
circulation of cooling water 冷却水循环
citrate, n. 柠檬酸盐
claim, n. 要求，声称；vt. 要求，声称
clarification, n. 澄清，说明
classical model 经典模型
climate, n. 气候，风土
climatic region 气候区域
clone, n. 克隆，无性繁殖；vt. 复制
close to 接近
cluster, n. 簇；vi. 群聚
coarse, adj. 粗糙的
coating, n. 涂层
cobalt, n. [化] 钴

cobaltiferous, adj. 含钴的
coil, n. 线圈，卷
colibacillus, n. 大肠杆菌
collaborate, vi. 合作，协作
colloid, n. [物化] 胶质；adj. [物化] 胶质
colloid chemistry [物化] 胶体化学
colonization, n. 殖民，移殖
colony, n. 菌落
column, n. 纵队，圆柱形物
column leaching, 柱浸
combination, n. 组合，联合
combine with 连同，联合
commission, n. 委员会
common, adj. 普通的；n. 公有地
compare to 与……相比
comparison, n. 对比，对照
competitive with 与……竞争
complexation reaction 络合反应，配位反应
complexes, n. 复合物
component, n. 成分，组件；adj. 组成的
compound, n. 化合物；vt. 合成；adj. 复合的
compressor, n. 压缩机
concentrate, n. 精选，富集，精矿
concentration, n. 浓度
conduct, vt. 管理，引导
configuration, n. 配置，组合，结构
confocal microscope 共焦显微镜
confuse, vt. 使混乱，使困惑
confusion, n. 混淆，混乱
conjunction, n. 结合
consensus, n. 一致，共识
consequently, adv. 因此，结果
consistency, n. 一致性，稳定性，稠度
constant, n. 常数，常量；adj. 不变的
constituent, n. 成分
construct, vt. 建立，建造；n. 概念
consumption, n. 消费，消耗
continual, adj. 持续不断的，频繁的
continual agitation 连续搅拌
continuous, adj. 连续的，持续的
continuous biooxidation 连续生物氧化
continuously, adv. 连续不断地

contribution, n. 贡献
controlling factor 控制因素
convection, n. 传送，对流
conventional, adj. 常规的，传统的
conventional process 常规流程；常规方法
conversion, n. 转换
convert into 使……转变，把……转变成
convert to 转换为
conveyor belt 传送带
conveyor, n. 传送机，传送带
cooling effect 冷却效应
copper, n. 铜；adj. 铜的
copper metallurgy 铜冶金学
copper precipitate 铜沉淀
copper product 铜材，铜制品
copper recovery 铜回收率
copper sulfide 硫化铜
copper sulphide ore 硫化铜矿石
copper-catalyzed bioleaching 铜催化浸出
correlate with 使有相互关系
correlation, n. 相关，关联
correspondence, n. 一致，相当
corrosion, n. 腐蚀
corrosion inhibitor 阻蚀剂
corrosion pattern 腐蚀模式
corrosion rate 腐蚀速率
cost-effective, adj. 划算的，成本效益好的
covellite, n. 靛铜矿
crawler, n. 履带牵引装置
cross sectional area 横截面积
crush, vt. 压碎；n. 粉碎
crystal lattice [晶体] 晶格点阵
curing time 固化时间，硫化时间
current density 电流密度
cyanidation, n. [无化] 氰化，[冶] 氰化法
cyanide, n. [无化] 氰化物
cycle time 周期
cytochrome, n. [生化] 细胞色素
cytoplasmic, adj. 细胞质

D

deal with 处理，对待

decade, n. 十年
decrease, vt. 减少，减小
defoamer, n. 去沫剂，消泡器
degrade, vt. 降级，降低
deliberate, adj. 深思熟虑的
density, n. 密度
dependence, n. 依赖，依靠
dependence on 依靠，依赖
deposit, n. 沉积物；vt. 使沉淀
derivation 引出
derivative, n. [化]衍生物，派生物
derive, vt. 引出，源自
descalant, n. 去垢剂
describe, vt. 描述，描绘
desulfovibrio desulfuricans 脱硫弧菌，去磺弧菌
detailed, adj. 详细的；v. 详细说明
dielectric constant [电] 介电常数
diffusion, n. 扩散
dimension, n. 尺寸
diminish, vi. 减少，缩小
dioxide, n. 二氧化物
dipole, n. [物化]偶极，双极子
direct mechanism 直接机制
direct oxidation 直接氧化
direct recovery 直接回收
disagreement, n. 不一致，争论
discharge, vi. 排放，放电
discovery, n. 发现
discuss, vt. 讨论，论述
disperse, vt. 分散，传播；adj. 分散的
dispersion, n. 散布，弥散，分散
dispersion rate 分散率
displacement, n. 取代，移位
displacing, v. 置换；adj. 置换的
disseminate, vt. 宣传，散布
disseminated, adj. 散播性的
dissolution, n. 分解，溶解
dissolution rate 溶解速率
dissolve, vt. 溶解，消失
distribution, n. 分布
disturb, v. 打扰，妨碍
disturbance, n. 干扰

disulfide, n. 二氧化硫
divide, vt. 划分
dominant role 主要角色
dominant specie 优势物种
dominate, vt. 控制，支配
donor, n. 捐赠人，供者
dosage, n. 剂量，用量
drain system 排水系统
drainage, n. 排水，引流
drainage piping 排水管
draw into 使卷入，摄取，吸收
driving force 驱动力
dross, n. 渣滓，浮渣，碎屑
due to 由于
dump, vt. 倾倒，倾泻
dump leaching 堆浸
duration, n. 持续
dust, n. 尘土，灰尘；vi. 化为粉末
dynamic, adj. 动态的，动力的；n. 动力
dynamic equilibrium [化]动态平衡

E

earthquake potential 地震的可能性
economic recovery 经济采收率
effect on 影响
effective, adj. 有效的，起作用的
efficiency, n. 效率，效能
electrochemical, adj. 电化学的
electrochemical oxidation [化]电化学氧化
electrochemical process 电化学过程
electrode, n. 电极
electrolysis n. [化] 电解
electrolyte solution [化]电解液
electrolytic dissolution 电解沉积
electron acceptor, 电子受体，电子接收体
electron, n. 电子
electronic, adj. 电子的
electrophoresis, n. [化] 电泳
electrophoretic mobility 电泳迁移率
electrostatic, adj. 静电的，静力学的
electrostatic repulsion [电] 静电排斥
electrowinning, n. [冶] 电解沉积，电解冶金法

elemental, adj. 基本的，主要的
eliminate, vt. 消除，排除
emerge, vt. 出现，浮现，摆脱
emergency, n. 紧急情况，突发事件
empirical, adj. 经验的，主观的
empirical constant 经验常数
empirical correlation 经验关系式
encapsulated, adj. 密封的
encounter, vt. 遭遇
energetically, adv. 积极地
enhance, vi. 提高，增强
enrichment, v. 富集
environmental pollution 环境污染
environmental protection 环境保护
enzymology, n. 酶化学
equation, n. 方程式，等式
equilibrium, n. 均衡，平静，平衡
equimolar, adj. 等物质的量的
equipment, n. 设备，装备
equivalent, n. 等价物，相等物
essentially, adv. 本质上；本来
establish, vt. 建立，建造
evaluate, vt. 评价，评估
evaporation, n. 蒸发，消失
evaporation rate 蒸发率
evaporative, adj. 蒸发的，成为蒸汽的
ever-increasing, adj. 不断增长的，持续增长的
evidence, n. 证据
evident, adj. 明显的
evolutionary, adj. 进化的，发展的
examine, vt. 检查
exception, n. 例外，异议
excessive, adj. 过多的，极度的
exert, vt. 运用，发挥
exiguous, adj. 稀少的，细小的
exist, vi. 存在
exothermic, adj. 发热的，放热的
expand, vi. 发展，展开
expel, vt. 驱逐，开除
experiment, n. 实验；vi. 尝试
explain, v. 解释，说明
exploit, vt. 开发，利用

extensive system 分支系统
extract, vi. 提取
extractant, n. 萃取剂
extraction, n. 抽取，取出
extractive, adj. 提取的，萃取的；n. 提取物

F

facilitate, vt. 促进，帮助
faculative, adj. 任意的，偶发的
Faraday's constant 法拉第常量
favorable, adj. 有利的，良好的
feasibility, n. 可行性，可能性
feature, n. 特征，特性
fermentor, n. 发酵罐，发酵桶
ferric, adj. 三价铁的
ferric arsenate [无化] 砷酸铁
ferric hydroxide 氢氧化铁
ferric ion 三价铁离子
ferric iron 三价铁
ferric sulfate 硫酸铁
ferric sulfide 硫化铁
ferrobacillus ferrooxidans 氧化亚铁铁杆菌
ferrous ion 亚铁离子
ferrous iron 二价铁
ferrous sulphate, [无化] 硫酸亚铁，硫酸铁
fertilizer, n. 肥料
fibril, n. 细纤维
filtration, n. 过滤，筛选
fine material 细粒材料
fissure, n. 裂缝；vt. 裂开，分裂
fixation, n. 固定，定位
flagella, n. 鞭毛，鞭节
flask, n. [分化]烧瓶，长颈瓶
floatation, n. [物化] 浮选
flocculant, n. 凝聚剂
flotation tailing 浮选尾矿
flue gas desulfurization gypsum 排烟脱硫石膏
fluid flow 液体流动
fluid loss 液体损失
fluorescent, adj. 荧光的；n. 荧光
fluorescent antibody 荧光抗体
for the purposes of 为了……目的

force, n. 力；vt. 促使，推动
formate, n. 甲酸盐，蚁酸盐
formation, n. 形成
fraction, n. 分数，部分
frequently, adv. 频繁地，时常
full-scale, adj. 全面的
function groups 官能团
fundamental, n. 基本原理；adj. 基本的
fungi, n. 真菌，菌类
fungus, n. 真菌
fungus mycelium 真菌菌丝
fungicide, n. 杀菌剂，杀真菌剂
furnace non-ferrous metallurgical 有色冶金炉

G

galena, n. [矿] 方铅矿，硫化铅
galvanic, adj. 电流的，使人震惊的
galvanic corrosion 接触腐蚀，[化工] 原电池腐蚀，[化] 电化学腐蚀
gangue, n. 脉石
garnet, n. [矿] 石榴石；adj. 暗红色的
gaseous emission 气体排放
gene, n. 基因，遗传因子
generator, n. 发电机，发生器
genus, n. 类，属
geometric, adj. 几何学的，几何的
geotechnical, adj. 地质技术的
geotechnical issue 岩土工程问题
glance at 瞥到
gluconate, n. [有化] 葡萄糖酸盐
goniometer, n. 测角仪
govern, vt. 控制，支配
gradual, adj. 逐渐的，平缓的
grain boundary 颗粒边界
grind, vt. 磨碎，研磨
grind size 研磨粒度
growth condition 生长条件
gypsum, n. 石膏

H

halide, adj. 卤化物的；n. 卤化物
hazardous, adj. 有危险的，有害的
heap, n. 堆，许多
heap leaching 堆浸
heat of reaction 反应热
hematite, n. 赤铁矿
hemihydrate, n. 半水化合物
heterogeneous, adj. 多相的，不均匀的
heterotroph, n. [生] 异养生物
heterotrophic, adj. [生] 异养的
heterotrophy, n. [生] 异养，异养生物
heterozygous, adj. 杂合的
hexadecane, n. [有化] 十六烷
hexahydrate, n. 六和水合物
high density polyethylene 高密度聚乙烯
high-grade, adj. 高级的，优质的
hillside, n. 山坡，山腰
humidity, n. 湿度，湿气
hybridization, n. 杂交，配种，[化] 杂化
hydrate, n. 水合物，氢氧化物；vt. 使成水化合物
hydrazine, n. [无化] 肼，联氨，酰肼
hydrocarbon, n. [有化] 碳氢化合物
hydrogen electrode 氢电极
hydrogenase, n. [生化] 氢化酶
hydrolysis, n. 水解作用
hydrometallurgy, n. [冶]湿法冶金
hydrophilic, adj. [化] 亲水的
hydrophobic, adj. 不易被水沾湿的，疏水的
hydrophobic interaction 疏水作用
hydroprocessing 加氢处理
hydroxide, n. 氢氧化合物

I

identical, adj. 同一的，完全相同的
identify, vt. 确定，认同
illustrated, v. 阐明
immunity, n. 免疫力
immunofluorescence, n. [免疫]免疫荧光
impeller, n. [机]叶轮
impermeable liner 防渗衬垫
impurity, n. 杂质
in addition 另外
in comparison to 与……相比
in conjunction with 连同，共同

in contrast　相反
in detail　详细地
in order　整齐，井然有序
in situ leaching　原位浸出
in situ　原位，原地
in some cases　在某些情况下
in terms of　根据，依据
in the absence of　缺乏，不存在
in the case of　在……情况下
in the form of　以……的形式
in the presence of　在面前
in the process of　在……的过程中
include, vt.　包含，包括
incorporate, vt.　吸收，包含；vi. 混合
incubation, n.　孵化；[病毒] 潜伏
indirect, adj.　间接的
indirect leaching mechanism　间接浸出机制
indirect mechanism　间接机制
indirect oxidation　间接氧化
industrial plant　工业厂房
inexpensive, adj.　便宜的
ingress, n.　进入，入口
inhibit, vt.　抑制，禁止
inhibition, n.　抑制，禁止
inhibitor, n.　抑制剂
inhibitory, adj.　禁止的，抑制的
initial, adj.　最初的
initial condition　初始条件
initial design　初始设计
initially, adv.　最初地
initiation, n.　启动，开始
injection well　注入井
inlet air　入气口
innovation, n.　创新，改革
inoculated, adj.　接种过的；v. 给……接种
inoculation, n.　接种，孕育
inoculum, n.　接种体，培养液
inorganic, adj.　无机的
installation, n.　安装，装置
integral, adj.　主要的；n. 部分，完整
integrate, adj.　整合的，完全的
intend as　打算把……当做

intend to　打算做，想要
interact with　与……相作用
interaction, n.　相互作用，[数] 交互作用；n. 互动
interchangeably, adv.　可交替地
interdependence, n.　相互依赖
interface, n.　界面，接触面
interference, n.　干扰，冲突
interfering, v.　妨碍；adj. 干涉的
intermediate species　中间物质
intermediate, n.　中间物，媒介；adj. 中间的
international tender　国际招标
interpret, vi.　解释
interpretation, n.　解释，翻译
inventory, n.　存货，库存
investigate, v.　研究，调查
investigation, n.　调查，研究
involve, v.　包含，牵涉
involvement, n.　参与，牵连
ion, n.　离子
ionic, adj.　离子的
iridium, n.　[化] 铱
iron bacteria　[微] 铁细菌
iron filing　铁粉
iron-oxidizing acidophile　铁氧化嗜酸细菌
iron-oxidizing bacterium (IOB)　铁氧化菌
irrespective of　不论，不考虑
irreversible, adj.　不可逆的
irrigation, n.　灌溉，冲洗
irrigation rate　沥淋率
irrigation solution　沥淋液
irrigation strategy　沥淋方法
irrigation temperature　沥淋温度
isoelectric point　等电位点
isolate, n.　隔离种群；vt. 使隔离
isothermal equation　等温方程式

J

jarosite, n.　[矿] 黄钾铁矾

K

kerosene, n.　煤油
Klebsiella, n.　克雷白氏杆菌属

kinetics, n. [力]动力学

L

laden, adj. 满载的，充满的；v. 装载
lag, n. 落后，延迟；vi. 滞后；adj. 最后的
lag phase 停滞阶段
lateritic, adj. 红壤的，含有红土的
lattice, n. 晶格，格子
leach, vt. 浸出，过滤，萃取；n. 过滤
leach cycle time 浸出周期
leach operation 浸出操作
leach pile 沥淋法回收矿物堆
leach rate 浸出率
leaching bacteria [化]浸矿细菌
leaching reaction 浸出反应
lead, n. 领导，铅，导线；vt. 致使，引导；adj. 带头的；最重要的
lead to 导致
leptospirillum ferrooxidans 亚铁钩端螺旋菌
liberate, vt. 解放，释放
lime, n. 石灰；adj. 绿黄色的
lime mud (LM) 白泥，石灰渣
limestone pulp 石灰浆
limestone, n. 石灰岩
limit to 限于
linear function 线性函数
lipopolysaccharide, n. [有化]脂多糖
liquid phase reaction 液相反应
liquor, n. 酒，溶液
lithium-ion 锂离子
living organism 生物体
lixiviant, n. 浸出剂，浸滤剂
low energy cost 低能量消耗
low-grade 低品位
low-grade ore 低品位矿

M

macroscopic, adj. 宏观的，肉眼可见的
magnesite, n. [矿]菱镁矿
magnesium, n. 镁
magnitude, n. 大小，量集
maintenance, n. 维护，维修
manganese, n. 锰

manifold, adj. 多方面的，各式各样的
mantle, n. 地幔，覆盖物；vt. 覆盖
margin, n. 边缘，利润；vt. 加边于
markedly, adv. 明显地，显著地
mass balance 质量平衡
mass transfer [热]质量传递，传质
mass transfer coefficient [化工]传质系数
material permeability 材料的渗透率
mathematical, adj. 数学的
mathematical model 数学模型
matrix, n. 矩阵，模型
matte, n. 冰铜
maximize, vi. 达到最大值
mechanical agitation 机械搅拌
mechanism, n. 机制，机理
membrane, n. 膜，薄膜
meniscus, n. [解剖]半月板，弯月面
mercury, n. 汞，水银
mesh, n. 网眼
mesophile, n. [微]中温菌；adj. [细菌]嗜温性的
mesophilic, adj. 亲中介态的，嗜常温的
metabolize, vi. 新陈代谢；vt. 使新陈代谢
metal sulfide 金属硫化物
metallurgy, n. 冶金
microaerophilic, adj. 微量需氧的
microbe, n. 细菌，微生物
microbial, adj. 微生物的
microbial degradation 微生物降解（作用）
microbiological leaching, 微生物浸出
microcolony, n. [微]小菌落，小集落
microflora, n. [微]微生物区系，[微]微生物群落
microorganism, n. [微]微生物，微小动植物，细菌
microscopic, adj. 微观的
microscopic immunofluorescence technique 免疫荧光显微技术
mild steel 低碳钢
mineral, n. 矿物；adj. 矿物的，矿质的
mineralization, n. 矿化，矿化作用
mineralogical, adj. 矿物学的
mineralogy, n. 矿物学
minimize, vi. 最小化
minimum inhibitory concentration (MIC) 最低抑菌浓度

mining wastes 采矿废物
mixed potential [物化] 混合电势
mixotrophic, adj. [生] 兼养的，混合营养的
model parameter 模型参数
moderate, adj. 稳健的，温和的
moderate thermophile, n. [生] 中等嗜热菌
moderately, adv. 适度地，中庸地
moisture, n. 水分，湿度
molybdate, n. [无化] 钼酸盐
molybdenum, n. 钼
monitoring, n. 监视，监控
morphology, n. 形态，形态学
mother liquid 母液
motile, adj. 能动的，活动的；n. 运动型
multiple, n. 倍数
multistage, adj. 多级的，多阶段的；vt. 使成多级
municipal solid waste incineration (MSWI) 城市垃圾焚烧
mycelium, n. 菌丝，菌丝体

N

negligible, n. 微不足道的，可以忽略的
neutral, adj. 中立的，中性的；n. 中立者，非彩色
neutralization, n. [化] 中和，中和作用
nickel, n. 镍；vt. 镀镍于
nickel citrate 柠檬酸镍水合物
nickel oxalate 草酸镍
nickel-cadmium battery [电] 镍铬电池
nickeliferous, adj. 含镍的
nitrogen, n. 氮气，氮
nitrospira, n. 硝化螺菌
nodularization, n. 球化
nonbonding, adj. 非键的
non-ferrous metallurgy 有色金属冶金
nonviable, adj. 不能存活的
norm, n. 规范，基准
noteworthy, adj. 值得注意的，显著的
nutrient supplementation 营养补充
nutritional, adj. 营养的，滋养的

O

observation, n. 观察

on a large scale 大规模的
once-throug, adj. 单程的，非循环的
ongoing, adj. 前进的，不间断的；n. 前进
operability, n. 可操作性
operating temperature 操作温度
opportunity, n. 时机，机会
optimal, adj. 最佳的，最理想的
optimization, n. 最佳化，最优化
optimize, vt. 使有化，使完善
optimum, n. 最适宜条件；adj. 最适宜的
optimum operating temperature 最佳工作温度
optimum rate 最优率
orbital, adj. 轨道的
ore constituent 矿物成分
ore deposits 矿床
ore leaching bacterial 浸矿菌
ore mineralogy 矿石矿物学
organic, adj. 有机的，根本的
organic acid 有机酸，有机酸类
organic nitrogen 有机态氮
organic solvent 有机溶剂
organism, n. 有机体，微生物
oscillating, adj. 震荡的
osmium, n. 锇
osmosis, n. 渗透，渗透性
overall, adj. 全部的，全体的
overburden, n. 表土
overflow, vi. 溢出，泛滥
overriding, adj. 高于一切的，最重要的
oxalate, n. 草酸，[有化] 草酸盐
oxidation, n. 氧化
oxidation behavior 氧化行为
oxidation chemistry of the process 氧化反应过程
oxidation kinetics [物化] 氧化动力学
oxidation mechanism 氧化机理
oxidation process 氧化过程
oxidation rate 氧化率
oxidation reaction 氧化反应
oxidation-reduction potential (ORP) 氧化–还原电位
oxide ore 氧化矿
oxidoreductase, n. 氧化还原酶
oxygen mass transfer rate 氧传质速率

P

palladium, n. 钯
paradox, n. 驳论，反论
parallel, adj. 平行的，类似的
parameter, n. 参数
participation, n. 参与，分享
particle, n. 颗粒
particle model 粒子模型
particle size 粒度
partitioning, n. 分割，分区；v. 把……分成部分
passive, adj. 被动的，消极的
patent, n. 专利权；adj. 专利的
pellet, n. 小球，球团矿
penetrate, v. 穿透，渗透
penicillium simplicissimum 简青霉
pentose, n. [有化] 戊糖
per a unit [物] 单位体积，单位容积
percolator, n. 过滤器
periodic, adj. 周期的
periplasm, n. [生] 周脂，外周胞质
permanent, adj. 永久的，永恒的
permeability, n. 渗透性，渗透率
petroleum, n. 石油
phase contrast 相位差
phase, n. 相
phosphate hydrazine 磷酸肼
phosphate, n. 磷酸盐
phosphorus, n. 磷
phreatic, adj. 潜水的，井的
phylogenetic, adj. 系统的，系统发生的
physicochemical, adj. 物理化学的
physiological, adj. 生理学的，生理的
pile, vt. 挤，堆积
pilot, adj. 试点的
pilot study 前期研究
pilot testing 引导测试，半工业化测试
pit, n. 矿井，矿坑
pitting current [冶] 局部腐蚀电流
pitting potential [冶] 局部腐蚀电位
placement temperature 放置温度
planar, adj. 平面的，二维的
planktonic, adj. 浮游生物的，浮游的
plastic liner 塑料衬套
plating, n. 电镀，镀层
platinum, n. 铂，白金
platinum-group metal 铂族金属
plow, n. 犁
polar, adj. 两极的，正好相反的；n. 极面
polarizability, n. 极化性，极化度
polarization, n. 极化
polarization current density 极化电流密度
polyethylene, n. 聚乙烯
polymer, n. 聚合物
polymer concrete 聚合物胶结混凝土
polymetallic, adj. 多金属的
polymetallic ore 多金属矿物
polysaccharide, n. 多糖，多聚糖
polystyrene, n. 聚苯乙烯
polythionate, n. 连多硫酸盐
pond, vt. 筑成池塘；n. 池塘
porphyry, n. [岩] 斑岩
portion, n. 部分，一份；vt. 分配
potassium, n. [化] 钾
potential, adj. 潜在的；n. 电势
power plant 发电厂
precaution, n. 预防，警惕
precipitate, v. 使沉淀
precipitation, n. 沉淀
preconditioning, n. 预处理条件
predict, vt. 预报，预言
predominance, n. 优势，卓越
preferential, adj. 优先的，选择的
preferential leaching 选择性浸出
preferential oxidation 选择性氧化
prerequisite, n. 先决条件
pressure oxidation 加压氧化
presumably, adv. 大概，推算起来
pretreat, vt. 预处理
pretreatment process 预处理过程
primary, adj. 主要的，初级的；n. 最主要者
primary element 原电池
primary mineral 原生矿物
principle, n. 原理，原则
prior, adj. 优先的
prior to 在……之前

problematic, adj. 有疑问的，不确定的
profound, adj. 深厚的，意义深远的
prokaryote, n. 原核生物
prolonged, adj. 延长的，拖延的
proportion, n. 比例，部分；vt. 使成比例
proportional to 与……成比例
protein, n. 蛋白质
protonation, n. 质子化
pseudomonas fluorescens 萤光假单胞菌，荧光假单胞杆菌
pseudomonas stutzeri 斯氏假单胞菌
pulp, n. 矿浆
pulp density 矿浆浓度
pump, n. 泵；vi. 抽水
purification, n. 净化，提纯，精致
purify, vt. 净化，使纯净
pyrite, n. [矿] 黄铁矿
pyrite breakdown 黄铁矿分解
pyrolusite, n. [矿] 软锰矿
pyrometallurgical process 火法冶金过程
pyrometallurgy, n. 火法冶金学
pyrrhotite, n. [矿] 磁黄铁矿

Q

quantitative, adj. 定量的
quantities of 许多
quantum, n. 量子论
quartz, n. 石英

R

radial flow impeller 径流式叶轮
radial flow turbine 径向流涡轮机
radioactive wastes [原子学] 放射性废物
radionuclide, n. [核] 放射性核素
raffinate, n. 萃余液
range, n. 范围
rate equation [物]速率方程
rate-determining step 限速步骤
reaction duration 反应持续时间
reaction kinetic 反应动力学
reaction path 反应途径
reactor system 反应系统

reagent, n. 试剂，反应物
rearrangement, n. 重新整理，重新排列
recovery, n. 恢复，复原
recovery boiler ash (RBA) 回收锅炉灰
recovery rate 恢复速率
redox, n. 氧化还原反应，[化] 氧化还原剂
redox potential 氧化还原电位
reductase, n. 还原酶
reduction, n. 还原，减少
reductive, n. 还原剂
refer to 涉及，参考
referred, v. 参考，查阅
refining, n. 精炼；vt. 精炼
refractory, adj. 难熔的，难治的；n. 耐火材料
regenerated, adj. 再生的；v. 再生
regeneration, n. [生] 再生，重生
regression, n. 回归，退化
regression line 回归线
regulatory, adj. 管理的，控制的
reintroduction, n. 再引入
related factor 相关因素
relative, n. 亲戚，相关物；adj. 相对的
release, vt. 释放，发射；n. 释放
rely on 依赖
remediate, adj. 治疗的
reoxidation, n. 再氧化
represent, v. 代表，表现
representing, v. 代表，表示
repulsion, n. 排斥，反驳
repulsive, adj. 排斥的
requirement, n. 要求，必要条件
residence, n. 住宅，居所
residues, n. 残留物，剩余物
resistance, n. 阻力，电阻
respectively, adv. 分别的，各自的
respiratory chain 呼吸链
retrieve, vt. 检索，恢复
reverse, vt. 倾倒，倒转；n. 相反
reverse circulation 反循环
reversible, adj. 可逆的，可翻转的
reversibly, adv. 可逆地
revolve, n. 旋转，循环

rhodium, n. 铑
ribulose, n. 核酮糖
roasting, v. 烤，煅烧；adj. 灼热的
rotating, adj. 旋转的；v. 旋转
rotating drum 转筒
rubber lining 橡胶衬里
rule of thumb 经验法则
run-of-mine 原矿
rupture, vt. 破裂，破坏
rusticyanin, n. 铜蓝蛋白
ruthenium, n. [化]钌

S

saturation concentration 饱和浓度
scrap, n. 残余物；vt. 废弃
scrub, vt. 擦洗，使净化
secondary smelting 二次冶炼
sedimentation, n. [化] 沉淀，沉降
segregation, n. 隔离，分离
semiautomatic, adj. 半自动的
semiconductor-electrochemical model 半导体电化学模型
sensible heat loss 显热损失
sensitive, adj. 敏感的
separate into 把……分成
sequence, vt. 排序
severe, adj. 严峻的，强烈的
shake flask 摇瓶
shear rate 切变速率
shear, vi. 剪切；n. 切变
shewanella putrefaciens 腐败希瓦菌，腐败希瓦氏菌
shrink, vi. 收缩，畏缩
siderophore, n. [无化] 铁载体
significance, n. 意义，重要性
significant, adj. 重大的，有效的，有意义的
significant difference 显著性差异
simplicity, n. 朴素，简易
simultaneous, adj. 同时的
single-phase 单相的
slight, adj. 轻微的，少量的
slope, n. 倾斜，斜坡
slurry, n. 泥浆，悬浮液

smelter waste 冶炼废弃物
sodium, n. 钠
sodium sulfate 硫酸钠
soil pollution 土壤污染
solar, adj. 太阳的，阳光的
solar radiation 太阳辐射
solids suspension 固体悬浮液
solubilization, n. 溶解，增溶
soluble copper sulfate 可溶性硫酸铜
soluble, adj. [化]可溶的；可溶解的
soluble ions 可溶性离子
solution, n. 溶液，溶解
solution concentration 溶液浓度
solution loop 溶液循环
solvent, n. [化]溶剂
solvent extraction 溶剂萃取，溶剂提取
sorption capacity 吸附能力
sparge, n. 喷雾；vt. 喷射
sparingly, adv. 节俭地，保守地
specification, n. 规格，说明书
spectrum, n. 光谱，频谱
speculate, v. 推测，推断
spent, adj. 耗尽了的，废的
sphalerite, n. [矿] 闪锌矿
spherical, adj. 球形的，球面的
spiral, n. 螺旋；adj. 螺旋的
sprinkler, n. 洒水器
stability constant 稳定常数
stabilize, vt. 使稳定，安定
stacking equipment 堆垛装置
staining technique 染色法；染色技术
stainless steel 不锈钢
start-up 启动
start-up operation 启动操作
State Environmental Protection Administration of China (SEPA) abbr. 国家环境保护总局
sterile, adj. 不育的，无菌的
sterile solution 无菌溶液
stirred-tank 搅拌槽
stirring, vt. 搅拌；adj. 活泼的
stirring intensity 搅拌强度

stirring rate 搅拌速率
stock bunkers 储矿槽
stoichiometry, n. 化学计量学
straight rod 直杆
straightforward, adj. 简单的，坦率的
strain, n. 张力，负担
stripping machine 剥离机
subculture, vt. 接种
subsequent, adj. 随后的，后来的
subsidence, n. 下沉，沉淀
substitute, v. 取代，代替；n. 代用品
substrate, n. 基片，基质
sucrose, n. [有化] 蔗糖
sufficiently, adv. 充足地，充分地
sulfate, n. 硫酸盐；vt. 使成硫酸盐；vi. 硫酸盐化
sulfate ion 硫酸盐离子
sulfide, n. 硫化物
sulfide ore 硫化矿
sulfolobus, n. 硫化叶菌
sulfur, n. 硫黄；vt. 用硫黄处理
sulfur compound 含硫化合物
sulfuric acid [无化] 硫酸
sulfur-oxidizing bacterium (SOB) 硫氧化菌
sulphide, n. 硫化物
sulphide ore 硫化矿物
sulphur, n. 硫黄，硫黄色；vt. 使硫化，用硫黄处理
sulphuric acid [无化] 硫酸
summarize, vt. 总结，概述
superficial gas velocity 表观气体流速
supergene, n. [地质]浅生矿床；adj. 表生的
supersaturate v. 使……过度饱和
supplementation, n. 补充，增补
surface area 表面积
surface coverage 表面覆盖度
surface tension 表面张力
surfactant, n. 表面活性剂；adj. 表面活性剂的
surprisingly, adv. 惊人的，不出意外的
surround, n. 围绕物；vt. 围绕，包围
surrounding area 周边地区
suspension, n. 悬浮
synthase, n. [生] 合成酶
synthetic, adj. 合成的，人造的；n. 合成物

synthetic media 合成培养基

T

take place 发生
tank leaching 槽浸出
taxonomic, adj. 分类学的
tension, n. 张力，拉力
terminology, n. 术语
that is to say 换言之
the number of 数目
theoretical, adj. 理论的
thermophile, n. [微] 高温菌；adj. 嗜热的
thermophilic, adj. 适温的，喜温的
thermostatically, adv. 恒温地
thiobacilli, n. 硫杆菌属，产硫酸杆菌
thiobacillus acidophilus 嗜酸硫杆菌
thiobacillus caldus 硫杆菌
thiobacillus coprinus 硫杆菌鬼伞属
thiobacillus ferrooxidans 氧化亚铁硫杆菌
thiobacillus thiooxidans 氧化硫硫杆菌
thiocyanate, n. [无化]硫氰酸盐
thiosulfate, n. [无化] 硫代硫酸盐
thiourea, n. [药] 硫脲
thumb, n. 拇指
titration, n. [分化] 滴定，滴定法
tolerance, n. 公差，宽容
toxic concentration 毒物浓度
toxicity, n. 毒性，毒力
toxicity test 毒性测试
traffic, n. 交通，运输；vi. 交易
transmission, n. 传输
transport, n. 运输，vt. 运输
truck, n. 卡车
tubing, n. 管子，管道
typical, adj. 典型的，特有的

U

unadapted, adj. 不适应的
unattached, adj. 独立的，未被捕获的
unattractive, adj. 不吸引人的，没有魅力的
underground, n. 地下；adj. 地下的
underground mining 地下采矿

underlying, adj. 潜在的，根本的；v. 为……的基础
undertaken, v. 从事，开始
unequivocally, adv. 明确地
uniformly, adv. 一致地
unpredictable, adj. 不可预知的，不定的
unpredictable behavior 不可预测行为
upstream, n. 上游；adj. 向上游的
uranium, n. [化] 铀
utilize, vt. 利用，运用

V

valence, n. 化合价
valence band [物化]价带
valley, n. 山谷，流域
valuable metals 有价金属
variables, n. 变量
variety, n. 多样，种类
vat leaching 桶式浸出
velocity, n. 速率，周转率
ventilation, n. 通风设备，空气流通
versatile, adj. 多才多艺的，通用的
vertical, adj. 垂直的；n. 垂直线
viable, adj. 可行的，能养活的
vicinity, n. 临近，近处

viewpoint, n. 观点，看法
viscosity, n. 黏度，黏性
visualize, vi. 呈现
volatile, adj. [化] 挥发性的，易挥发的；n. 挥发物

W

warm climate 温暖气候
washout, n. 冲刷，失败者
waste pile 矸石堆，废石堆
waste, n. 浪费，废物；vt. 浪费；adj. 废弃的
welded, adj. 焊接的，焊的；v. 焊接
wetting agent 润湿剂
wind velocity 风速
with respect to 关于

Y

yield, vt. 出产；n. 产量，收益

Z

zinc, vt. 镀锌于……，涂锌于……，用锌处理；n. 锌
zinc sulfide [无化] 硫化锌
zone, n. 地区，地带；vt. 使分成地带
zoology, n. 生态

References

[1] RAWLINGS D E. Biomining: Theory Microbes and Industrial Process [M]. Springer, 1997.

[2] ROSSI G. Biohydrometallurgy [M]. McGraw-Hill, 1990.

[3] EHRLICH H, BRIERLEY C. Microbial Mineral Recovery [M]. Mcgraw- Hill, 1990.

[4] 杨显万, 沈庆峰, 郭玉霞. 微生物湿法冶金 [M]. 北京: 冶金工业出版社, 2003.

[5] 杨洪英, 杨立. 细菌冶金学 [M]. 北京: 化学工业出版社, 2006.

[6] BRIERLEY J A. A perspective on developments in biohydrometallurgy [J]. Hydrometallurgy, 2008, 94: 2~7.

[7] EHRLICH H L. Past, present and future of biohydrometallurgy [J]. Hydrometallurgy, 2001, 59: 127~134.

[8] BRIERLEY J A, BRIERLEY C L, Present and future commercial applications of biohydrometallurgy [J]. Hydrometallurgy, 2001, 59: 233~239.

[9] POULIN R, LAWRENCE R W. Economic and environmental niches of biohydrometallurgy [J]. Minerals Engineering, 1996, 9(8): 799~810.

[10] JAIN N, SHARMA D K. Biohydrometallurgy for nonsulfidic mnerals —A review [J]. Geomicrobiology Journal, 2004, 21(3): 135~144.

[11] WATLING H R. The bioleaching of sulphide minerals with emphasison copper sulphides —A review [J]. Hydrometallurgy, 2006, 84: 81~108.

[12] McDONALD R G, WHITTINGTON B I. Atmospheric acid leaching of nickel laterites review. Part II. Chloride and bio-technologies [J]. Hydrometallurgy, 2008, 91: 56~69.

[13] NDLOVU S. Biohydrometallurgy for sustainable development in the African minerals industry [J]. Hydrometallurgy, 2008, 91: 20~27.

[14] WATLING H R, PERROT F A, SHIERS D W. Comparison of selected characteristics of *Sulfobacillus* species and review of their occurrence in acidic and bioleaching environments [J]. Hydrometallurgy, 2008, 93: 57~65.

[15] PRADHAN N, NATHSARMA K C, SRINIVASA RAO K, et al. Heap bioleaching of chalcopyrite: A review [J]. Minerals Engineering, 2008, 21: 355~365.

[16] GERICKE M, NEALE J W, van STADEN P J. A Mintek perspective of the past 25 years in minerals bioleaching [J]. The Journal of the Southern African Institute of Mining and Metallurgy, 2009, 109(10): 567~585.

[17] HOQUE M E, PHILIP O J. Biotechnological recovery of heavy metals from secondary sources— An overview [J]. Materials Science and Engineering, 2011, C 31: 57~66.

[18] DAS A P, SUKLA L B, PRADHAN N, et al. Manganese biomining: A review [J]. Bioresource Technology, 2011, 102: 7381~7387.

[19] ANJUM F, SHAHID M, AKCIL A. Biohydrometallurgy techniques of low grade ores: A review on black shale [J]. Hydrometallurgy, 2012, 117~118: 1~12.

[20] ZENG G, DENG X, LUO S, et al. A copper-catalyzed bioleaching process for enhancement of cobalt dissolution from spent lithium-ion batteries [J]. Journal of Hazardous Materials, 2012, 199~200: 164~169.

[21] SHI S, FANG Z, NI J, et al. Comparative study on the bioleaching of zinc sulphides [J]. Process Biochemistry, 2006, 41: 438~446.

[22] PRADHAN D, PATRA A K, KIM D, et al. A novel sequential process of bioleaching and chemical leaching for dissolving Ni, V, and Mo from spent petroleum refinery catalyst [J]. Hydrometallurgy, 2013, 131~132: 114~119.

[23] SAMPSON M I, PHILLIPS C V, BLAKE R C, et al. Influence of the attachment of acidophilic bacteria during the oxidation of mineral sulfides [J]. Minerals Engineering, 2000, 13(4): 373~389.

[24] CRUZA R, LÁZARO I, GONZÁLEZ I, et al. Acid dissolution influences bacterial attachment and oxidation of arsenopyrite [J]. Minerals Engineering, 2005, 18: 1024~1031.

[25] AGHAZADEH M, ZAKERI A, BAFGHI M S. Modeling and optimization of surface quality of copper deposits recovered from brass scrap by direct electrowinning [J]. Hydrometallurgy, 2012, 111~112: 103~110.

[26] AHMADI A, SCHAFFIE M, PETERSEN J, et al. Conventional and electrochemical bioleaching of chalcopyrite concentrates by moderately thermophilic bacteria at high pulp density [J]. Hydrometallurgy, 2011, 106: 84~92.

[27] HANSFORD G S, VARGAS T. Chemical and electrochemical basis of bioleaching processes [J]. Hydrometallurgy, 2001, 59: 135~145.

[28] Behera S K, PANDA P P, SINGH S, et al. Study on reaction mechanism of bioleaching of nickel and cobalt from lateritic chromite overburdens [J]. International Biodeterioration & Biodegradation, 2011, 65: 1035~1042.

[29] GU G, SUN X, HU K, et al. Electrochemical oxidation behavior of pyrite bioleaching by *Acidthiobacillus ferrooxidans* [J]. Transactions Nonferrous Metals Society of China, 2012, 22: 1250~1254.

冶金工业出版社部分图书推荐

书 名	定价(元)
硫化铜矿的生物冶金	56.00
微生物湿法冶金	33.00
材料的生物腐蚀与防护	28.00
含砷难处理金矿石生物氧化工艺及应用	20.00
沼气发酵检测技术	18.00
生物柴油检测技术	22.00
有色冶金原理	35.00
有色金属冶金学	48.00
有色冶金概论	30.00
冶金专业英语	28.00
专业英语教程——材料成型与控制工程	26.00
金属材料成形双语教程	28.00
材料科学与工程专业英语精读	39.00
实用有色金属科技日语教程	33.00
连铸技术(全英文版)	25.00
现代流体力学的冶金应用(英文版)	25.00
Mechanical Properties of Materials	30.00
英汉冶金工业词典(修订本)	138.00
汉英英汉连续铸钢词典	65.00
大学实用英语语法	28.00
大学英语语法精要	20.00
大学英语写作与修辞	36.00
简明大学英语语法	29.00
简明实用英语写作	23.00
现代英语词汇学	19.50
英语教学理论探讨与实践应用	22.00
当代英语阅读教学论	26.00